WHAT PEOPLE ARE SAYING ABOUT *DIGITAL INC.*

"A stimulating and in-depth chronicle of publishing's digital revolution. . . . Publishing professionals will find this fascinating."—*Publishers Weekly* (starred review)

"An engaging narrative of an industry embroiled in unpredictable change . . . lively and definitive."—*Kirkus Reviews*

"Curtis's sharp observations illuminate questions that still echo today . . . Readers with an interest in publishing, content creation, or the future of media will find this a thoughtful, practical case study of real-time literary innovation."
—*Publishers Weekly BookLife*

"*Digital Inc.* is really astonishing. I'm sure no one else will ever attempt what Curtis has done here, giving the entire history of the e-book with this level of detail. There are details that are great fun and sentences worth savoring throughout. It deserves a serious place on the physical (and digital) shelf."—Bob Miller, CEO of Phaidon Press and former President of Flatiron Books

"A fascinating and informative trip through perhaps the most transformative period in book publishing history. When the gun went off to start the digital revolution, Richard Curtis was at the starting line!"—Jack Romanos, former CEO of Simon & Schuster

"For anyone who lived the e-book revolution in publishing, this is a must-read. For anyone concerned with how book publishing may cope with future changes for which it is never ready, Richard Curtis's account is an essential study."
—Donald Maass, literary agent and author of *The Emotional Craft of Fiction*

"The book publishing industry has changed more since the mid 1990s than it had since Gutenberg's invention of the printing press. *Digital Inc.* is an entertaining, informative, and accurate account of this tumultuous time from the perspective of an agile and successful participant."—John Ingram, Chairman, Ingram Content Group

"In *Digital Inc.*, Richard Curtis has penned the Rosetta Stone for anyone interested in the tectonic shift in book publishing from print to digital. Curtis not only tells a fascinating and extraordinarily well written business story but also shows how the dramatic changes in publishing impact our daily digital lives. I learned more from *Digital Inc.* than from any other book about the digital revolution."—Roger Cooper, former Senior VP and Publisher, the Berkley Publishing Group

DIGITAL INC.

DIGITAL INC.

From Print to E-Book—
Inside the Transformation
of the Book Industry

RICHARD CURTIS

Rivertowns
BOOKS
IRVINGTON, NEW YORK

Hardcover edition ISBN-13: 978-1-953943-72-9
Paperback edition ISBN-13: 978-1-953943-73-6
Ebook edition ISBN-13: 978-1-953943-74-3

LCCN Imprint Name: Rivertowns Books
Library of Congress Control Number: 2025944312

Rivertowns Books are available from all bookshops, other stores that carry books, online retailers, and directly from the publisher. Visit our website at www.rivertownsbooks.com. Retailer and consumer orders, inquiries about discounts for bulk purchases, and other correspondence may be addressed to:

Rivertowns Books
240 Locust Lane
Irvington NY 10533
Email: info@rivertownsbooks.com

For Leslie

". . . and you're the only one who knows."

Bliss was it in that dawn to be alive.
WILLIAM WORDSWORTH

CONTENTS

INTRODUCTION

ALTHOUGH MUCH HAS BEEN WRITTEN about the Digital Revolution, I'm not aware that the turbulent transition of books—and authors—from analog to digital has been recounted by an insider who actually participated in the revolution. The transformation occurred over a relatively short time span, a few decades, but they were jam-packed with astounding scientific wonders and creative breakthroughs. We take most of them for granted today, but in the late twentieth century and early in this one, the disparate elements were scattered like pieces of a fiendishly complex engine awaiting a master builder to assemble and integrate them into the publishing process. In this book, I have set out to describe the confluence of three different cultures—technological, business and legal—and my personal involvement as literary agent, e-book publisher and author.

This upheaval has been characterized as "digital disruption," and there was plenty of that as the venerable book industry clashed with a completely new means of production and distribution. But these battles were also fountainheads of innovation, and countless authors owe their careers and fortunes to the scientists and researchers who, invention by invention, brought forth the products we call e-books. The stakes were high: the future of reading (and arguably, writing).

The e-book revolution is also a perfect lens through which to examine the media revolution driven by artificial intelligence. The fervor intoxicating advocates of AI today is strikingly reminiscent of the zeal of many idealists who attended the birth of e-books. Both share the credo that liberating books from the strictures of copyright law will shower blessings on a benighted world. The slew of AI-related infringement lawsuits in progress duplicates the clashes waged at the dawn of the digital book era, right down to haggling over the word "snippets."

DIGITAL INC.

It has been a decade since I departed the e-fray, long enough to digest and gain perspective on the evolution of books and the people who write them from analog to digital. Even those who lived and worked in the e-book era, roughly 1980 to 2020, may be too close to the experience to be fully aware of how their world shifted beneath their feet and how they themselves evolved from one mindset to another. I hope this story will help them see it all more clearly.

Richard Curtis

January 2026

PRELUDE: APPOINTMENT IN GAITHERSBURG

(1998)

But are our values turning asswards
When opening books requires passwords?[*]

I N 1998, AT THE DAWN of the digital book era, while still the head of a literary agency, I started an e-book publishing venture. Or should I say "adventure"? Though not exactly an expedition to the crocodile-infested headwaters of the Nile—indeed, my journey was largely sedentary—it was nevertheless bold, exhilarating, and even perilous. I was not young in 1998, but I'm ready to swear I am in greater possession of my wits today than I was when I launched that company, for I was blissfully ignorant of the risks and obstacles that lay ahead. Looking back, I wonder how I got through it with a measure of success; indeed, how I got through it at all.

The story actually starts thirteen years earlier, on a fine spring day in 1985. As I walked from the subway to my office on East 64th Street, listening to music

---

[*]*Author's note:* In 1985, I began composing end-of-year poems for *Publishers Weekly* magazine, offering comments on notable publishing people, events and trends of the preceding twelve months. Excerpts from these verses are offered at the start of each chapter of this book, representing my at-the-moment impressions of the unfolding history of our industry. I gave up the practice in 2024, feeling all rhymed-out after decades of brain-wracking annual versification.

on my Sony Walkman, an image suddenly flashed in my mind's eye. Instead of inserting a music cassette in the device, what if I could insert one to read?

I shunted the image aside when I arrived at my office. Yet, in idle moments that day, this epiphany flared anew, evoking fantasies of texts projected onto the screens of portable paperback-size devices. The image lodged itself in my brain, and as time went by, I came across news stories that researchers were actually working on such a device. I strongly felt the allure of this emerging technology, and every report of progress intensified it. Over the next thirteen years, as I tracked developments, I grew more and more certain the scenario I had imagined was going to become a reality.

I was therefore intrigued to receive an invitation to a conference on October 8 and 9, 1998, called Electronic Book '98 Workshop, sponsored by the National Institute of Standards and Technology (NIST), in Gaithersburg, Maryland, a suburb of Washington DC. The subtitle of the event, "Turning a New Page in Knowledge Management," sounded rather academic, so I was totally unprepared for the buzz of excitement when I entered the elegant wood-paneled Green Auditorium. Starry-eyed scientists, technologists and publishing people filled every one of the room's 293 seats, forcing another hundred to stand at the back of the room.[1]

Although the convocation was billed hyperbolically as "the first of its kind," there had been other such events before this one. Just eight months earlier, a conference called Book Tech '98, held in New York City's Marriott Marquis Hotel, drew some 2,600 publishing professionals to 1,400 booths displaying the latest advances. Yet that expo did not generate the galvanic charge that animated this event, perhaps because this one was hosted by the U.S. government, suggesting Uncle Sam's official blessing over the proceedings.[2]

Three recent game-changing breakthroughs enhanced the excitement of the attendees. The first was the rollout of the earliest commercial e-book readers, the Rocket eBook and Softbook. The second was the introduction at Chicago's Book Expo, in the summer of 1998, of a print-on-demand machine that enabled users to feed a digital text file into a printer at one end and a printed book would come out of the other. The third was the realization that the principal source of texts to feed e-books was not cassettes, floppy disks or ROM cards; it was the Internet.

The air crackled with the evangelical zeal of true believers who had been inspired by a vision similar to my own. As I talked to them, it became clear that for many, this was the realization of a decades-old dream. They had long toiled in narrow and obscure corners of research and development, and this was the first time they would be able to see the larger picture and hear or meet researchers whose work dovetailed with their own. Now, at last, their labors would achieve recognition and acclaim.

I had often speculated on what role I might play when this medium finally came to fruition. For reasons I couldn't express, the traditional literary agent's model of licensing rights for a commission didn't satisfy me. I wanted to be more proactive but didn't yet know what that would look like. Maybe I would find the answer here.

I scanned the audience but found no familiar face. The attendees seemed to be divided between tech people casually dressed in jeans and sport shirts, and publishing people in business attire. The former outnumbered the latter by a serious multiple, and the predominant feeling was definitely geeky.

The contrast in clothing belied a deeper clash of cultures. As I circulated among the conferees, I could sense the deep divide. Whereas publishing people felt that books were the finest means of communicating information ever created, the tech element saw them as the doomed artifacts of another era. One contingent envisioned a revolutionary new medium; the other's attitude was "If it ain't broke, don't fix it."

I had to sympathize with the revolutionaries. Although computerization had manifested itself in various editorial and production aspects of book publishing, in 1998 the fundamental process of taking a text from paper manuscript to bound, printed volume was not vastly different from what it had been in the early twentieth century or, for that matter, in the early eighteenth. Editorial relationships were established via lunches, drink dates and parties, and business negotiations conducted on landline phones. Manuscripts were submitted by mail or messenger, contracts and checks transmitted via the U.S. postal system. Paper copies of contracts, correspondence and reviews were stored in metal file cabinets.

So many book production procedures were performed manually, from copyediting with pencil on paper to production of galley proofs, from cover paintings to catalogs. Editor and publishing consultant Stephen S. Power, reminiscing about his employment at that time, recounts how "I personally took the script around to every department so that I had face time with everyone involved in the process, from jacket copy to cover art to tip sheets to sales to marketing to publicity."

I studied the convocation's program and attendee list. Speaking, serving on panels or simply observing the proceedings (taking notes on laptops) were delegates from such prominent corporations as Intel, Motorola, 3M, Xerox, Sony, Lucent Technologies, Hewlett Packard, Hitachi, Kodak, Minolta, IBM and Microsoft. The list was filled with e-book developers sporting sexy futuristic names: Nuvomedia, Autotext, Netopolis, Librius, TeleRead, Pixelworks, Electric Press, Glassbook, Softbook, Cybook, Everybook and Hardshell Word Factory.

Interspersed with them (taking notes with pens on paper) were people from the publishing sector ranging from Simon & Schuster in New York to Editora Atica in Brazil to Kadokawa Shoten in Japan; also librarians from New York City's Columbia University and from as far away as Alaska, and educators from California, Wisconsin, North Carolina and Maryland, among them, numerous librarians and representatives of organizations for the blind and physically handicapped. And listed as well were some whose connection to e-books seemed remote: Procter & Gamble? The U.S. Supreme Court? The Naval Surface Warfare Center? Boeing Flight Safety?

A few attendees had feet in both worlds, men and women from the technology sector engaged by publishers to translate technical jargon for the hidebound world of book publishing. (The day when every big publisher had a "Chief Digital Officer" in the executive suite was still over the horizon.)

Strangely—almost bizarrely—no one from Amazon attended despite its boast of being "Earth's biggest bookstore." A word search through NIST's proceedings did not even produce a mention of the company's name or representative from the behemoth.

Some other absences were also conspicuous. No one from OverDrive, Inc., which was on its way to building a gargantuan network of digital-book lending libraries, was on the roster. Nor was anyone from Franklin Electronics, which had been producing handheld electronic reference books, dictionaries, translators and Bibles for over a decade. And though the name of Gutenberg was frequently invoked at the convocation, no one was there from Project Gutenberg, which had pioneered digital commerce more than twenty-five years earlier by means of public domain texts delivered via a primitive precursor of the Internet. (There was, however, an attendee named Jonathan Guttenberg, President of New Media at Random House.)

Despite the nonattendance of these key players, the anticipation of the throng was sky high. When NIST Director Raymond Kammer took the podium and declared, "Welcome to the first Electronic Book Workshop!" the audience erupted in cheers. The man next to me said, "This is the biggest moment since 1455."[3]

The presentations began, and everyone sat patiently through talk after talk wondering where they fit in and how they could make money in this emerging technology. Many were struggling with nomenclature. Today we all know what we mean when we refer to electronic books. But it was far from clear that everyone in the room shared a common understanding or (except for terms like "Lunch in the Cafeteria") even spoke the same language. Book people grappled with terminology like "microencapsulated electrophoretic display," "electrowetting" and "plasmonic nanostructures." Was "64-bit Random Access Memory" a lot or a little? Was a gigabyte a little or a lot?

The geek element struggled to understand publishingspeak terms like "trim size," "fair use" and "hard-soft deals." Did "sub rights" have something to do with fighting ships? Was "pub date" a rendezvous where editors went for drinks with agents?

In this Babel-like atmosphere, the application of "digital" to "publishing" threw familiar terms into a cocked hat. If authors were now able to sell and distribute their work directly to readers—as this evolving technology empowered them to do—didn't that make them publishers? If publishers created original, copyrightable digital content, didn't that make them authors? If a literary work took the form of a digital file delivered electronically, was it still a book?

Hypertext, the nonlinear computerized language developed in the 1960s, raised bewildering issues for publishing people. If you could not only choose your own ending to a book but the beginning and middle as well, what was a story? And if, using hypermedia to link to video and audio, you could (in science fiction author Greg Bear's term) "type a movie"—well, what then was a movie? Did that make authors auteurs?

During a break I chatted with an engineer type. "What do you for a living?" he asked. I said, "I'm a literary agent." He looked puzzled. "I represent authors," I explained. Another blank stare. "I sell their manuscripts to publishers. I negotiate their book contracts."

He brightened and snapped his fingers. "Ah! You're a content provider!"

The head-spinning complexity of the issues was exemplified in a number of talks about paper and ink. Not your grandparents' paper and ink—electronic paper and ink. The future of reading was no longer bound sheaths of paper but texts on screen. We were introduced to a number of ingenious variations on paper, such as lightweight plastic sheets that you could literally turn like the leaves of a book. All these products employed plastic or glass sheets in which a matrix of crystals, beads or transistors was embedded or suspended in liquid. When charged by an electric current, the beads in the matrix arranged themselves as texts or images.

Most promising was what MIT researchers called E Ink, which displayed texts and images on screens clearly and rapidly. There was an undercurrent of sniggering in the audience as a straight-faced scientist explained how an electric current could turn your white balls black and your black balls white. (MIT had the last laugh when E Ink was sold to Prime View International, a Taiwan company, for $215 million.)[4]

All joking aside, electronic paper and E Ink were clearly essential to the future of e-books. But making them as legible and convenient as their analog prototypes—well, easier said than done.

Take a traditional book—what many publishing people smilingly call a "book book" to distinguish it from its electronic counterpart. The pages are bright white. You can read them in sunlight. You can read them in the shade. You can read them in candlelight. Graphs, charts and photos are easily inserted adjacent to the text referring to them. Books come not just in black-and-white but in every color of the rainbow. When you turn a page, there is no lag while you wait for the next to appear. If you accidentally drop it on the sidewalk, it doesn't shatter or cease operating. When you put it down, you can save your place by inserting a bookmark or (if you are a barbarian) dog-earing the page. And if you close it for a day, a week or a decade, it will start up instantaneously and function exactly as it did when you last put it on the shelf.

Best of all? Book books don't require batteries.

We hold those truths to be self-evident. But for scientists and engineers, reproducing the look, feel, flexibility and durability of book books was a maddening challenge. Some screens, such as those in tablet computers using LCDs (liquid crystal displays) glared blindingly in sunlight. On the other hand, LCD screens, which are backlit, could be read in the dark; E Ink screens then in development could not. LCDs could display color; E Ink could not.

A big deal was screen angle. You can read a printed book even if you hold it nearly horizontally, but the state of screen technology in 1998 made it hard to read the text unless you held the screen perpendicularly.

Pagination was another huge headache: When you enlarged the font, the text on page 50 became page 80. A caption that went with a photo on page 100 now shot to page 125 if you enlarged the font or to page 75 if you reduced it. If you had to turn your screen from portrait to landscape in order to see an illustration, the graphic flipped back to portrait, requiring you to tilt your head 90 degrees to make the image out.

And then there was that damn refresh rate, the time it took for the text to appear when you clicked to the next page. The delay was annoyingly, unacceptably long, like the lag in some early digital cameras between the press of the finger and the click of the shutter.

Dan Munyan, CEO of Everybook, Inc., summed it up neatly: "What a computer does best, a book cannot do, but what a book does best, a computer cannot."

One more thing: On a dedicated e-book reader you couldn't make a phone call, watch a movie, play a videogame or listen to music, applications long visualized by science fiction writers. Consumers yearning for a handheld device incorporating audiovisual media and telephonic capability would profoundly affect the future industry. But today, at NIST, the focus was on the elusively simple task of reproducing the experience of reading a printed book.

To laypeople in the audience, the presentations were not easy to follow. Some of the papers and PowerPoint demonstrations were abstruse, complex and boring. And so, when the talk about batteries was announced, the audience buckled down for another tedious presentation. They knew of course that the battery was (to borrow Tracy Kidder's phrase) the soul of this new machine, an indispensable component of any portable electronic device. But it was not going to be easy to endure a lecture devoted to "phase-controlled metal-oxide-based materials for improved energy storage."

Then the Battery Lady stepped up to the podium and magic happened.

Svelte, beautiful, charming, a trained opera singer, Swedish scientist Dr. Christina Lampe-Önnerud had been a dedicated researcher in the battery field since 1991 and had invented a number of innovative techniques (today she holds over 80 patents and a superabundance of honors).[5] Her talk alliteratively focused on five factors dictating the future of batteries, all of which were still wanting at this point in time: "Coverage, Convenience, Clarity, Control and Cost." Though the pages of her PowerPoint presentation seem dry and monotonous to read, her artless and soft-spoken delivery, with the faintest of Swedish accents, mesmerized the crowd. One observer quipped that her enunciation of "lithium ion" was as seductive as Eartha Kitt's rendition of "Santa Baby."

As the conference wound up on the second day, a significant truth became manifest: Despite the messianic zeal of the participants and breathtaking progress on numerous scientific fronts, e-books simply weren't ready for prime time. Aside from shortcomings in the devices' science and technology, few standards had been developed on which everyone could agree. Many fields of research and development were in competition or not ready to integrate with each other, harking back to such standards battles as the videotape format war between the Betamax video cassette recorders (VCR) and VHS (Video Home Systems) in the 1970s and '80s.

Lack of standards and the absence of a clear business model plagued the embryonic e-book industry. How much would consumers pay for an e-book reader? How much would they pay to download a book? When, after print publication, should the e-book edition be released? If you bought a printed book, did you have to pay extra for the e-book? How much royalty would publishers pay to authors? What means would be used to load digital texts into e-books? Floppy disks? Smart cards? Internet? Could you enhance an e-book with audio, photos and videos? Could you embed links to a dictionary and thesaurus, to other books, to movies? How could you maintain control over the public dissemination of digital files? How could you protect your properties against piracy?

A number of essential publishing channels were being neglected. Julia Blixrud, Senior Program Officer of the Association of Research Libraries, described some fiendishly challenging issues in digitizing, formatting, cataloging and circulating library-compatible e-books. Another source of turmoil was books for the blind, dyslexic and handicapped. Judith M. Dixon, PhD, Consumer Relations Officer at the National Library Service for the Blind and Physically Handicapped, Library of Congress, reported how desperately e-books were needed for some forty-four million visually impaired people forty-five or older, four million of whom were severely afflicted. The only available resources were in Braille or on records and cassettes, and only three to five percent of print books and one half of one percent of magazines were produced in recorded or Braille formats. Electronic media, including Braille-speech combinations, could be an enormous force for good—if the people at this conference would only apply themselves to developing the technology.

Perhaps the most important issue of all was almost entirely overlooked. Because the thrust of the conference was technological, the matter of copyright scarcely had the same impact as touch screens, E Ink and battery life. Indeed, there seemed to be a pervasive air of ignorance, indifference or insouciance about the subject among the attendees. Carol Risher, Vice President of

Copyright and New Technology Association of American Publishers, recognized this truth in a cautionary speech. "In the enthusiasm to consider the wealth and variety of literature and information products that will be available to import to e-book devices, there hasn't been much mention of the issues raised by copyright—ownership, control, protection against copying, licensing, derivative works, authenticity."

The significance of Risher's talk didn't seem to register, as evidenced by one speaker who gushed that the new technology would enable users to scan all of the world's literature and make it available to the masses free of charge. At a subsequent convocation I felt compelled to stand up and say something:

> It must come as painful news to those of you who have grown up believing that if you see a piece of text you like, you just hit the "Copy" command, paste it, and forward it to all your friends on the Net. Unlike us folks from the content sector who jealously protect the sanctity of intellectual property, many of you grew up with a strong sense of entitlement to it during the freewheeling early days of the Internet. And many of you still believe content should be given away free— unless of course you created it, in which case you will sue the ass off anybody infringing your work.
>
> So, I can't blame you technical people for fuzzy thinking about content. I can only remind you that a book is protected by copyright law the moment it comes out of an author's brain and it is fixed on a medium like paper, and until and unless that author assigns or licenses the rights to another party, the law prohibits reproduction or dissemination without the owner's express permission. Failure to do so is called infringement, a legal term meaning you can't do it.

Publishers and authors controlled a stupendous storehouse of copyrighted content—books, stories, articles. But with so many unanswered questions, so much confusion and contradictory information, so many loose ends flapping in the breeze, the possessors of this treasure were reluctant to venture out with it. Until perfect clarity in all these matters was achieved, the realization of a thriving e-book industry would be stalled.

Given the fact that the conference was sponsored by the National Institute of Standards and Technology, a number of presentations focused on the imperative need to develop standards. The abstract of one by Steve Stone, Director of Electronic Books Development at Microsoft, read:

> For electronic books to succeed, tens of thousands of titles must be
> available immediately upon release of the electronic book devices.
> Publishers are willing to provide content if a standard file format and
> content structure exists into which they can write their content as
> well as a digital rights management system that can protect their con-
> tent. Companies that understand this will work quickly and aggres-
> sively to correct this by developing electronic book standards with
> the correct content structure as well as content protection features.

In his closing remarks, Victor McCrary, Technical Manager of the Information Storage and Interconnect Systems Team in NIST's Information Technology Laboratory, described the road map for a successful e-book industry and
called for committees to come back the following year with a set of universal
standards that would put everyone on the same page (pun intended) in such
matters as device weight, power requirements, content format, display reflec-
tivity, voice recognition, content security and integrity—and above all, software
platform. Only when the end users—readers—were satisfied would the industry
take off. That meant a device that looked and felt like a book: portable and lightweight, with pages bright in light or darkness; possessing such software features
as dictionary, highlighting and font resizing; low power consumption; capable
of carrying illustrations; and "multi-path distribution of content—wire, wireless
or portable media."

Two years earlier, Microsoft cofounder Bill Gates had written an influen-
tial article entitled "Content Is King."[6] But at the NIST conference, the throne
was occupied not by content but by hardware. The realization of the e-book
dream would be postponed for years while countless issues were sorted out.

Though everyone who attended still believed that e-books were inevitable,
e-books obviously weren't arriving tomorrow morning, and the conference ad-
journed beneath a cloud of uncertainty. The bottom line? No standards, no business model; no business model, no business.

In short, the challenges facing the nascent industry in 1998 could not be
more daunting. It was a precarious time to be an e-book publisher.

By the time I boarded the train for New York I had made up my mind to
become one.

1. ANALOG AGENT

(THE 1980s)

What Nostradamus could foresee
The wonder known as POD?
Could Gutenberg in his vainglory
Imagine life sans inventory?

T HE VISION I EXPERIENCED that day in 1985 may have been a revelation
to me, but as I researched the topic over the next dozen years, I learned
that it was far from original. In fact, not only had people dreamed about e-books
but someone had actually built a viable prototype more than fifty years earlier.
Around 1930, a visionary named Bob Brown invented a device for projecting
books onto a screen. He called them Readies (pronounced "*reed*-ies," the literary equivalent of "movies").[1]

See if this description, written seventy-seven years before the Kindle,
sounds familiar:

> Extracting the dainty reading roll from its pill box container, the
> reader slips it smoothly into its slot in the machine, sets the speed
> regulator, turns on the electric current, and the whole 100,000,
> 200,000, 300,000 or million words spill out before his eyes . . . in one
> continuous line of type. . . . My machine is equipped with controls so
> the reading record can be turned back or shot ahead . . . [including a]
> magnifying glass . . . moved nearer or farther from the type, so the
> reader may browse in 6 point, 8, 10, 12, 16 or a size that suits him.

More recently, in the 1960s and '70s, academic and military researchers had developed content delivery models that (in principle, at least) closely resembled today's e-book, including the ability to upload plain text via ARPANET, a rudimentary version of the Internet developed by the military. Project Gutenberg, founded in 1971, was a collection of public-domain works in digital format distributed over the ARPANET. It might serve as a model for a commercial publisher (except that its books were given away free, a condition at which commercial publishers might balk).

These early applications required enormous computing power or other advanced technological resources. In 1985, genuinely portable readers were way down the road, far too remote for me to do anything practical with the idea. It was a pleasurable daydream, but I had a business to run.

The business I had to run was the literary agency I started in the 1970s, which illicitly occupied several rooms of an Upper East Side brownstone converted into offices. Widespread disregard of New York City's law prohibiting businesses above the first floor of brownstones (except doctors and lawyers) was one of the poorest kept secrets in town, but as long as we paid rent to our landlord and taxes to our city, we were left in peace.

Thanks to our growing prosperity, we had recently modernized our operation with a touch-tone and voicemail phone system, electric typewriters, LaserJet printer, Xerox photocopier and fax machine.

Our staff included myself, a junior agent, a bookkeeper and "the Kid"—a trainee who sorted the mail, packed submissions, shlepped packages to the post office and ran errands.

Our agency's stock in trade was genre fiction, highly commercial, formulaic novels issued as mass market paperback originals in such categories as westerns, romances, thrillers, science fiction, fantasy and horror.

My involvement with genre fiction was purely serendipitous. An American Studies major and editor of *The Syracuse Review*, the university's literary magazine, I graduated college with dreams of following in the footsteps of my idol Henry James, complete with velvet smoking jacket, soirees at country estates and witty repartee in elegant ateliers of prominent socialites. Unfortunately, I needed a job, and there were no listings at the employment agency for men of letters in velvet smoking jackets.

There was one, however, for a position with Scott Meredith Literary Agency, its founder a tough and shrewd wheeler-dealer whose agency specialized in genre fiction appealing to male readers. The agency boasted a stable of top science fiction writers including Arthur C. Clarke, J. G. Ballard, Philip K. Dick, and detective novel authors Richard S. Prather, Evan Hunter ("Ed McBain") and Donald Westlake ("Richard Stark"). As Scott's reputation grew, he attracted some significant literary lions, notably Norman Mailer.

My immediate boss was Scott's brother Sidney, a businessman of pedestrian competence, who needed someone to organize a department devoted to handling foreign rights, a potentially lucrative source of revenue. As I literally did not know what rights were, I'm not sure what special acumen he saw in me. "The first thing we need for you to do," Sidney explained on my first day, "is to figure out how to open these goddamn aerograms without tearing them in half." Aerograms were prestamped sheets of thin paper on which messages were written, folded in thirds and sealed, a clever alternative to costly international airmail letters. He showed me numerous examples that he had sliced in twain with his letter opener and taped back together. I mastered the requisite skills, learned what rights were and went on, over time, to build the foreign rights department into a profit center. When I left the firm years later, I typed up a procedural manual to guide my successors. Sidney had it laminated and, for years afterwards, whenever I bumped into veterans of Meredith's foreign desk, they spoke reverently of "The Manual" as if it were a grimoire of spells and incantations.

In time, I was permitted to handle some science fiction, mystery and other genre novelists. My original impression when I joined the agency had been that this literature was as far from *belles lettres* as comic strips. My condescension was inexcusable; I had never read any of that stuff. Now it was my job to do so, and the experience was transformative. The more I read, the more I recognized how dedicated, disciplined and skillful these storytellers were, and I came to admire and bond with this remarkable cadre of artisans, the best of whose works—I will stand on this assertion—rivals those of many authors of mainstream fiction.

After seven years in Meredith's service, I left to become a freelance writer. I'd published some mystery stories plus six or seven racy novels for an outfit in Chicago. I make no apologies; they were a great way to learn how to write fast and well, and many leading novelists got their start writing them. One of the best things about this fiction mill is that the books went straight to the printer without editorial review. One day I asked the publisher why he didn't employ an editor. "Let me ask you, kid," he replied, "what the fuck do you need editors for?" From time to time, I have pondered that question.

Thanks to my apprenticeship in the darker byways of the writing trade, I developed into a fairly facile writer and launched a full-time career. Over the next ten years, I published dozens of works of fiction and nonfiction, and I suppose that had I stuck with it "I coulda been a contenda." But around this time, a former Meredith client asked me to handle his work, and I made some deals for him. I actually had two desks in a tiny office I rented: one for my writing, with my beloved Airedale Fumfer snoozing with her muzzle on my right foot; the other for agenting, with Fumfer snoozing on my left one.

The lure of a profession where I could make more from the commission on someone else's book than I could from selling one of my own pulled me away from authorship. I named my first company The One Horse Literary Agency, but eventually other fugitives from Meredith cast their lot with me and I abandoned One Horse in favor of Richard Curtis Associates, Inc., incorporated in 1979.

Paperback originals were a nursery for talented newcomers who might write a dozen books that broke even or lost money for their publishers before their gifts were recognized by fans and rewarded with tolerable compensation. Some of them developed into bestselling mainstream authors. The market for this literature was robust: Readers were ravenous and authors prolific. Paperback houses issued scores of category books every month written by writers capable of delivering three or four or more books a year. Genre fiction made money, and when I started my own agency, that became my specialty. Although hardcovers may have been more profitable per unit, the huge sales of mass market paperbacks compensated for the absence of prestigious literati in our stable. Even with high return rates, the economics were profitable. You could make a good living handling science fiction, romance, thrillers and horror, and we did.

A stroke of fate gave our fledgling agency's fortunes a strong boost. An agent friend invited me to share offices with her, and we found a space on East 52nd Street in midtown Manhattan. As we were moving in, I noticed the name on the office next door: Robert P. Mills, Ltd. Bob Mills was a former science fiction magazine editor and founder of a highly respected literary agency with a sterling list of crime and science fiction clients. Our offices were separated by one wall! From time to time I bumped into Bob, a lanky and slightly stooped gentleman, and shared pleasantries with him in the elevator. That casual acquaintance was altered by an incident that occurred one evening in 1980.

My soon-to-be wife Leslie and I attended a science fiction banquet. The master of ceremonies, science fiction author Norman Spinrad, was endeavoring to rhapsodize about NASA's achievements but was being heckled by some character at the front of the room. Actually, *harassed* is a better description of the snarky interruptions and insults leveled at Spinrad by this loudmouth. No, *invaded* is the most accurate word of all, for this rude fellow climbed up on the dais, took over the lectern and unleashed a rant.

"Who is that?" Leslie asked.

"That's Harlan Ellison," I said. "He's one of the leading voices in fantasy and science fiction."

"He's certainly one of the loudest," she said.

As we were leaving, who should walk up to us but Ellison himself! "Are you Richard Curtis? I've heard good things about you. I gotta see you."

"Sure!" I handed him my card.

He stared at it. "It's the same address as my agent."

"I know," I said.

He came to my office the next day. "I gotta leave Bob Mills," he said in a hushed tone, as if afraid Mills could hear through the wall that separated us. "He hasn't done shit for me."

We discussed representation. He seemed satisfied with what I had to offer and we shook hands. Whereupon he left my office, went next door and, after a short while, returned from the Mills office with a pile of cartons filled with his books and files.

In the next year or two, a few more dissatisfied Mills clients left him for our agency, compelling me to send a note to Bob inviting him to lunch. I thought he might like to get out before he lost all of his authors. After a drink or two I said, "Bob, would you be interested in selling your agency to me?"

He took a few months to ponder the question and then reached out to me. "Only because my wife is sick and I must take care of her," he explained.

We worked out terms and I acquired his agency, tripling our client list and, after I closed his office, our revenues. His wife passed away soon afterwards, and Bob himself in 1986.

The fabled agent Marvin Josephson said, "The relationship of an agent to a publisher is that of a knife to the throat." My former boss Scott Meredith used to say, "People respect you more if you take something away from them than if you give them something." But though I was aggressive on behalf of my clients, ruthlessness was never my modus operandi. I preferred the Speak Softly and Carry a Big Stick approach. If you represent a list of important clients, you don't need to pound the table and shout to win a negotiation. All it takes is to say something like, "Are you sure you want to do that?" to make a wise editor reconsider a demand. One of my authors nicknamed me the Velvet Shark.

Many of my authors expressed interest in the publishing process and enjoyed hearing my explanations about how the business worked. That prompted me to start "Agent's Corner" early in the 1980s, a kind of "inside publishing" column for *Locus*, a science fiction trade magazine. My pieces were designed to educate authors about industry practices, ranging from a clause-by-clause analysis of a book contract to interpreting royalty statements to navigating movie deals. I also offered less formal guidance, such as how to read option clauses (they're made to be broken), what happens at a sales conference (editors and agents jump into the swimming pool with their clothes on), and how authors should behave at a publishing lunch (don't suck on lobster claws). In time, I addressed the advent of electronic publishing.[*]

Some columns focused on serious issues and inequities, about which the author community seemed surprisingly uninformed. Even more notably, after the column was picked up by other book trade publications, I learned that many editors who followed me were in the dark about those issues, too, and I received many a note from them saying, "Gee, I never knew that!"

One thing that nobody seemed to know anything about was how royalties were calculated. In theory, the computation should be straightforward enough: In a royalty deal, authors are supposed to receive a percentage of the sale of every copy of their books. To figure out your royalty, you multiply the list or net price of your book by the percentage defined in your publishing contract. The product of your calculation should equal the amount written on your royalty check. In actuality, the royalties paid to you (if and when they were paid to you) seemed to have no correspondence whatever to this simple formula. And neither I nor my agent colleagues could reckon how the publishers arrived at the figures in their royalty statements, which were invariably lower than expected.

That may be forgivable for authors, but for agents? Even many editors were uncertain about it, too, perhaps because they were not always privy to royalty data generated by their company's royalty accounting department. All

---

[*] These columns were collected into a number of books, the first of which was *How to Be Your Own Literary Agent*, Houghton Mifflin, 1983; revised in 1984, 1996, and 2003.

too often, when authors questioned or complained about royalty statements, their editors were in the dark.

I set out to educate myself, and some of the things I learned were alarming. Which explains why, in the early 1980s, I embarked on a crusade.

The object of my campaign was consignment bookselling, a practice initiated during the Great Depression that enables book stores to return unsold copies to publishers for full credit. In a piece called "Flight to Quantity" I described it thus:

> Despite decades of proof that returnability is a sucker's game, the publishing industry is incapable of curing its addiction to the practice.
>
> The time has come for publishers to accept the fact, now glaringly apparent to all but those in total denial, that no business enterprise can afford to sell just half or even two-thirds of what it manufactures—*and* to foot the bill for the return and disposal of the unsold stock. Some pundits ascribe the woes of our business to printed books themselves, saying that the medium is no longer appropriate for our times. In truth nothing is wrong with printed books. Everything is wrong with the way they are distributed.
>
> And the way they are distributed is appallingly profligate, taking a dreadful toll on the environment in terms of paper waste and carbon footprints. The tortuous methods by which bookstores account to publishers and publishers to authors are imbecilic and arguably fraudulent. An alien visitor tracking the journey of a printed book today—from editorial office to printer to warehouse to bookstore, back to warehouse and then to remainder jobbers or pulpers—would have genuine reason to wonder whether there is intelligent life on this planet.

It was not just this foolish and wasteful practice, almost unique among industries, that provoked my scorn, but the perverse royalty accounting system built on it. The publishers' chain of reasoning went like this: "If we do not know how much money we may have to refund to booksellers returning unsold stock, we cannot know how much royalty to pay to authors. We will therefore hold onto their royalties until returns are finalized."

This withholding of royalties, known as "reserves against returns," was potentially abusive because publishers could—as many did—hold excessive amounts of royalties for intolerable lengths of time. (When my son was born, a client suggested I name him Royalty Statement, because he was late and smaller than expected.)[2]

In another blog article, "The Decline and Fall of the Mass Market Paperback," I explained:

> Royalty reports to authors were deliberately fashioned to omit information about the number of copies printed, shipped, and returned, or about the amounts of royalty reserved pending finalization of returns. This suppressing of vital sales data gave publishers carte blanche to retain royalties that should have been remitted to authors. Some publishers got too creative and held royalties forever. Until the 1990s, when pressure from agents and writers' organizations forced publishers to reveal significant details, mass-market houses reported only net sales with no information as to how they arrived at those net figures. It was like reporting batting averages to baseball fans without revealing how many at-bats or hits the players had.

Other than browbeating editors and royalty managers, there were few ways to obtain or verify the calculations behind the royalty check you (sometimes) got. An editor told me of a publisher he worked for who had stacks of unsigned royalty checks on his desk. "When are you planning to send these out?" my friend asked. "When the authors sue me," the publisher replied.

For an agent to publicly take publishers to task is risky if not foolhardy, and my big gadfly mouth finally got me into hot water. One day I got a call from Ron Busch, president of mass market publishing for Simon & Schuster. "A word to the wise," he said. "If you persist in haranguing us about our business practices, you may find it difficult to sell books to us."

You didn't have to be particularly astute to grasp his warning, and I realized that it might be the better part of prudence to stop biting the hands that fed me and my authors. But—miraculously—the gods intervened. Not long after Busch's threat, I got a call from Bill Dailey, husband and business manager of Janet Dailey, one of the most successful romance writers in America. "We've been reading your column and really admire your guts," he said. "We need someone aggressive like you to represent us." Shortly thereafter Janet signed with our agency.

And who was her publisher?

Simon & Schuster.

I called Ron Busch. "Hey, Ron, guess what?" I chirped. "I am now Janet Dailey's agent."

My heart thundered as I waited for a torrent of wrath. Busch's fulminations were legendary.

At last, he spoke. "Let's have lunch."

He took me to a four-star restaurant.

Bill Dailey was a shrewd and plain-spoken westerner, small in stature but oversized in ambition for his wife's career. The legend goes that on their honeymoon, she looked up from the saccharine romance she was reading and said, "I could write better books than this."

"Well," he said, "get off your butt and write one." She did, launching a career that took her to the top of her profession. Some years later, I helped her recover dozens of her romances from Harlequin and put them into my e-book program.

When Bill Dailey passed away, I flew out to Missouri to attend his funeral, a joyous country-and-western-music wingding. The night before, I visited the funeral home where he lay in an open coffin, dressed in jeans, plaid shirt, and boots. Hooked to his belt was his ever-present cell phone. A blinking red light indicated that—Omigod! It was on! What if someone tried to call him? I approached the funeral director. "Do you see Bill's cell phone is on?"

"I know," he said.

"Why?" I asked.

"In case," he replied.

Aside from my broadsides, I was actually in a position to exert some influence on publishers, having been elected president of the newly formed Independent Literary Agents Association. I had also been engaged by the Science Fiction Writers of America to serve as its advocate for grievances. The latter organization was particularly proactive, bringing pressure on publishers to make their royalty reporting more transparent. We even audited one publisher. Today, every legitimate publisher provides information about copies reserved against returns, plus projected dates when withheld reserves will be released.

The consignment distribution model, however, remains in place to this day.[3] Although pundits don't cite it as a principal cause of the shrinking of viable publishers from hundreds to a handful, the fact is that only megapublishers can afford to carry a business that loses a serious percentage of revenue to returned merchandise. Big houses hemorrhage the same as small ones; it just takes longer for their lifeblood to drain.

Surely there had to be a way to distribute books that was not dependent on returns. And there was—decades away. Nobody could see it in the 1980s, but when it did come it would upend every assumption publishers cherished. I focused on building my agency, but kept tracking digital technology. That blip of insight I had experienced in 1985 while listening to my Walkman grew brighter and brighter, a beacon lighting my path to unknown worlds.

2. THE DREAM OF PORTABILITY

(1989)

Now the decade's end draws nigh.
Our battle-weary eyes are dry.
We bid farewell without a pang
To beaucoup Sturm and plus de Drang.
It avails us naught to rake over
Tales of ten grim years of takeover
And multinational aggression
That doomed the Gentleman's Profession.

A s THE 1980s PROGRESSED, advances in e-book-related computer research and development came thick and fast:

- Breakthroughs in personal computing (IBM's PC in 1981 and the Apple Macintosh in 1984)
- The advent of compact discs and, in 1982, CD-ROM storage
- Word processing programs like WordStar (1978), WordPerfect (1979), Microsoft Word (1983) and MacWrite (1984)
- The rise of desktop publishing
- Refinement of powerful object-oriented programming languages like C++ (1985)
- Introduction of laptops
- Dramatic developments in microprocessing, circuit integration and miniaturization

- Numerous other leaps in hardware and software. By 1990 the storage capacity of computer hard drives had topped one gigabyte[1]

Could there be any doubt where all this was heading?

Yet the means for stuffing all that content and functionality into a palm-size book-reading device remained elusive. Reducing the size and weight of the device, miniaturizing its operating system, scaling down power consumption, extending display time—and designing software to manage all of these elements—these were among the many obstinate elements that had not been pacified. Moore's Law, the prediction made in 1965 by Intel Corporation founder Gordon Moore that computer efficiency would double every two years, did not yet govern e-book research and development.

Until 1989, when Burlington, New Jersey, home of Franklin Electronics, brought forth the Bible.

Since the late '70s computer scientist Peter Yianilos, founder of a company called Proximity Technologies, had been exploring a number of word processing functions and devices. He reasoned that if you could hold a mathematical calculator in your hand you could hold an electronic dictionary or thesaurus. He writes that in 1988, after his company merged with Franklin Electronics, his research "formed the basis for the first hand-held electronic books, ranging from spellers and dictionaries to Bibles and encyclopedias."[2] He held patents in a number of products that his work had fashioned. His gadgets sold in the millions. (Steve Jobs and engineer Michael Hawley had explored some of the same challenges in the mid-1980s and ended up embedding a dictionary, thesaurus and book of quotations in their NeXT computers.)[3]

They were self-contained devices with dynamic search and display tools, arguably making the Franklin Bible the first true e-book. Its first version (KJ-21, for "King James 21st century"), weighing in at 1,125 megabytes of read-only memory, held the complete English texts of all 66 books of both the Old and New Testaments totaling 1,888 pages, and incorporated phonetic pronunciation, spelling lookup, footnotes, bookmark function, database search and retrieval, data compression, speech recognition and even a help key. It employed a physical keyboard and—you are permitted to smile here—an LCD display of four lines of text. The device was a triumph of compression, squeezing 4.5 megabytes of text down to less than 1 megabyte.[4] It weighed well under a pound: 11.8 ounces, excluding four double-A batteries.

2. THE DREAM OF PORTABILITY

For e-book visionaries, the Franklin Bible was a harbinger: today Genesis, tomorrow *War and Peace*. And yet, as exciting and innovative as it was, the device was still primitive. You could not fit it into pocket or purse. The keyboard was difficult to navigate, resolution poor, print tiny and screen hard to read in daylight. More important, its storage capacity was limited. Even 1,888 pages was a drop in the ocean of available content. And the device lacked a simple and efficient means for uploading the vast amounts of other material in the world. The Internet would of course one day become *the* purveyor of unlimited content, but in 1989 that was four years away. Users were stuck with manually feeding texts on floppy disks or ROM cartridges into desktop computers tethered to wall outlets.

Rudimentary as the Franklin Bible was by today's standards, it was the state of the art at the end of 1989 and good enough to hearten the growing ranks of believers, some of whom may have read *Cyberbook*, a perceptive novel by sci-fi author Ben Bova about the impact of e-books on a futuristic publishing world, that was published the same year.

The impact of the Franklin Bible was not limited to devotees of the Good Book. Some publishing executives had also seen it and were pondering how they could turn the technology to their advantage, as I rudely found out one morning in 1992 with my discovery of a UCO—an Unidentified Contractual Object.

I had recently made a deal with Berkley, a mass market paperback publisher owned by Putnam Publishers (now a division of Penguin Random House). As I reviewed the contract boilerplate, I came across a provision I had never seen. It was titled "Display Rights." As a close reader of publishing contracts, I felt a surge of anxiety, like an astronomer who has just detected an ominous celestial body entering the solar system.

The provision granted to Berkley the right

> to display the Work in any manner designed to be read and to license the display of the Work in any manner designed to be read, in whole or in part, by any means, method, device or process sequentially or nonsequentially ("Display Rights"), including without limitation mechanical visual recordings or reproductions (together with accompanying sounds), and all other forms of copying or recording of the Author's words and/or illustrations in any manner designed to be read, which are not either granted to the Publisher elsewhere in this agreement or reserved to the Author.

I called Phyllis Grann, Putnam's formidable publisher. "What's the story with Display Rights, Phyllis?"

"Well," she explained, "I went to an electronics trade show and saw this gadget"—she described the Franklin Bible—"and decided we want the rights to books that can do that."

"You realize you're seizing control of electronic rights," I protested.

"Yes," she said matter-of-factly.

"How negotiable are you on this clause?" I asked.

"We're not," she answered.

This was very troubling. Putnam's clause raised some fundamental questions plus a host of lesser but nevertheless significant ones, all of them disturbing. The biggest one was where on the publishing spectrum did e-books sit?

There were three ways to look at it. The first might be called "A Book Is a Book Is a Book." That is, no matter what the format or how it is delivered to the reader— a bookstore shelf, a computer screen, hieroglyphs on a clay tablet—the texts are identical. What difference does the label on the bottle make if it carries the same wine? Granting a publisher the right to print your book on paper, went the argument, implies the right to display the same text on a screen.

The second way to characterize e-books was to view them as a subsidiary right, an inherent component of a publisher's basic contract package. If your publisher has the absolute right to license your book to a book club, doesn't it have the absolute right to license it to an e-book reprinter?

The third explanation was the most intriguing: It posited that e-books constitute an entirely new and distinctive medium, as different from a printed book as a movie or an audiobook, and therefore exempt from publishers' traditional suite of entitlements. That meant that in order for your print publisher to issue an e-book edition, they would have to create specific language in your contract, including a royalty schedule and a percentage of licensing revenue.

No matter how you defined it, Phyllis Grann's "Display Rights" was the locus where the x axis of analog publishing crossed the y of digital.

The fact that Putnam had to create new contractual language to sew up e-book rights to future books suggested they were not sure they possessed the rights to previously published ones. That sent publishers, authors and agents scurrying to their contract files to pore over agreements seeking language indicating that they did or did not control e-book rights to their books. What they found (when they found them at all, for many were decades old) was a hodgepodge of vague, confusing or contradictory phraseology. But the essential language seemed to boil down to a few essential phrases.

One was "information storage and retrieval," which suggested that at the time the contract was issued, there existed some sort of automated method of retaining and recovering a book's text. That did not mean—as one publisher liberally interpreted it—sending the intern down to the storage room to retrieve a carton of photostats or microfiches (an early form of photocopied documents). It meant an electronic process —or a "mechanical" one, in the words of Putnam's clause—and its existence in an old book contract strongly supported a publisher's claim of control over e-rights.

Another significant term everyone sought as they scoured old contracts was anything sounding like "now in existence or hereafter devised." This wording declared in effect, "Right now, the only way to store text may be Xerox copies, but in case someone comes up with a bigger and better storage and retrieval system, we hereby assert that it falls under our control."[*]

Then there was the Competitive Works clause, which prohibited authors from publishing or authorizing any work that was substantially similar to the published book and was likely to injure the book's sales. As this provision was, and still is, in the boilerplate of just about every book contract, it offered protection to publishers from versions (such as e-books) that contained all or much of the text of the original work.

Finally, they looked for something along the lines of "All rights not granted are reserved to the Author." That would be bad news for the publisher.

These linguistic issues were not merely theoretical. They became casus belli in a growing number of hassles, not a few of which were bitterly fought; some ended up in court. A groundbreaking case arose when members of the National Writers Union, an alliance of some 27,000 authors and journalists founded in 1981 by Jonathan Tasini, sued a number of periodicals including *The New York Times, Newsday* and Time Inc. for copyright infringement. These publishers had licensed the writers' works to electronic databases without

---

[*] In light of developments in AI, a contemporary version of this provision might read: "Any new technology, applications, media, formats, mode of transmission, and/or methods of distribution, dissemination, exhibition, or performance now existing or developed in the future, are deemed granted and conveyed to the Publisher."

permission and without compensating the writers. The case went all the way up to the Supreme Court, where, in 2001, the court confirmed that the plaintiffs' work had been infringed Justice Ruth Bader Ginsburg delivered the 7–2 opinion, resulting in an award of $18 million to the writers. The lesson for publishers was make sure your contracts with authors are explicit about what rights you're buying and what you're going to pay for them.

An even higher-profile case had the potential to make law. It happens that Arthur Klebanoff, an attorney who had started an e-book publishing company called RosettaBooks in 2000, had made deals with authors to publish e-book editions of a number of Random House books. The list included such significant works as *Cat's Cradle* and *Slaughterhouse-Five* by Kurt Vonnegut, Jr., and *Sophie's Choice* and *The Confessions of Nat Turner* by William Styron. Original publication dates of these books ranged from the 1950s to the 1970s. They were still in print. That is, printed copies were still on sale according to the publisher's definition of "in print."

Klebanoff had examined Random's contracts. They simply stated that Random had the right to publish these works "in book form." As there was no such thing as "e-book form" when the contracts were drafted, Klebanoff reasoned that Random's ownership did not include e-rights. He pointed out an analogous situation: after mass market paperbacks ("pocketbooks" as they were then termed) were introduced in 1939, publishers had amended their contracts to ensure they legally controlled those rights. Their new language defined pocketbooks as a separate format and specified a royalty schedule. Did not the same principle apply to this new format called e-books? In the absence of express language and a schedule of royalties covering e-rights, Klebanoff asserted, Random House did not own them.

No sooner had RosettaBooks announced its plans to release the e-books than Random House leaped into action, filing a complaint that Rosetta was infringing their rights. They also requested a federal injunction to block Rosetta's release of the books.

The issues were not just big, they were stupendous. As bizarre as it may seem, there was no definitive legal definition of the artifact known as a book. From Gutenberg to modern times, the device of choice for displaying works designed to be read sequentially, and whose content could be stored and retrieved, was called a book. Random House, using the "A Book Is a Book Is a Book" rationale, argued that e-books are simply versions of print editions and that it was legally sufficient for a contract to state that Random would publish the literary work "in book form." This was by no means a disingenuous argument: After all, wasn't the text on a screen identical to the text on a printed page? That was Random's contention.

Given how much was riding on the case, other publishers like Simon & Schuster and Penguin Putnam fell in behind Random House, while the Authors Guild and the Association of Authors' Representatives backed Klebanoff.

The verdict sent shock waves through the industry: Random House's request for an injunction was turned down by U.S. District Judge Sidney H. Stein.

Random House appealed the decision.

Given the expense (a subsequent suit, *HarperCollins v. Open Road*, cost $1.5 million), publishing lawsuits inevitably end up getting settled unless there is a principle so utterly critical that the parties are willing to fight to the death. And this one, *Random House, Inc., Plaintiff-appellant, v. Rosetta Books LLC and Arthur M. Klebanoff, in His Individual Capacity and As Principal of Rosetta Books LLC, Defendants-appellees*, 283 F.3d 490 (2d Cir. 2002), was such a case. The appeal was heard by the U.S. Court of Appeals for the Second Circuit.

Random House lost again.[5]

"Without expressing any view as to the ultimate merits of the case," the decision stated, "the Court concludes that the district court did not abuse its discretion in denying Random House's motion for a preliminary injunction, and consequently the judgment is affirmed."[6] In other words, the issue was not whether Klebanoff had infringed; it was whether Random House could exercise prior restraint of his e-publications. The court said they could not. If Random House wanted satisfaction, they would have to wait until the e-books were published. *Then* they could sue for infringement.

The judge recommended they try to settle, and with risk and expense mounting for both parties, they decided that was a good idea. "Under the agreement," Hillel Italie wrote in *The Intelligencer*, in exchange for Rosetta's promise not to solicit any more Random House books, "Random House will grant Rosetta exclusive e-book rights to 'mutually agreed-upon titles,' both old and recently published."

And so, to this day, the question of what is a book remains unanswered.

In a subsequent case, Open Road Integrated Media, an e-book publisher founded in 2009, issued an e-book edition of Jean Craighead George's 1971 young-adult novel *Julie of the Wolves*. HarperCollins, whose traditional edition was still in print, sued and, in 2004, won. Harper's contract contained the phrase "now known or hereafter invented," which, the court concluded, was "sufficiently broad to draw within its ambit e-book publication."[7]

But that was not the end of mischief caused by the introduction of contractual provisions like Display Rights. This clause in Putnam's contract, "including without limitation mechanical visual recordings or reproductions (together with accompanying sounds)," introduced some extremely disturbing scenarios. For it just so happens that a mechanical visual recording with accompanying sounds is called a movie.

Not surprisingly, motion picture and television studios and agents, practically frothing at the mouth, jumped into the fray, warning authors and agents that if they granted that language to a publisher, it would kill a movie or television deal. A related medium was video games, which by the 1980s had become a red-hot market, as PC and arcade products made by Atari, Nintendo and Sega explosively proliferated. In 1989 the first handheld game player, Nintendo's Game Boy, was unleashed on the world, selling in the millions and creating a generation of young users who thought nothing of accessing entertainment on handheld devices.[8]

Even the seemingly innocuous phrase in the Display Rights provision, "together with accompanying sounds," was a land mine with the potential to blow up beneath every medium. To eagle-eyed lawyers it conjured all kinds of enhancements like multicharacter readings of novels that impinged (if not infringed) on performance rights and the evolving field of audiobooks.

On these issues of multimedia applications of e-books, the warring parties reached a détente of sorts by agreeing on a bifurcated e-rights clause for book contracts. Part A covered *verbatim* rights, the right to reproduce the exact text of a book as an e-book without any bells, whistles, tunes or other enhancements. This was a must-have for authors and agents, and the movie people were okay with it. Part B covered *adaptations* of the kind described in Putnam's Display Rights clause, language that was poison to movie and television producers. Publishers could retain Part A in their contracts but delete Part B, thus avoiding collision with movie, television, videogame and audio deals.

But what about that provocative word in the Putnam clause, "nonsequentially"? Short of peeking at the ending of a mystery novel, was there any other way to read a book but sequentially? In point of fact, there was. We knew it as interactive fiction of the "Choose Your Own Adventure" type, where readers elect any of a number of alternative story lines and endings. Not only were most children familiar with them in the 1990s, but they loved navigating the various pathways, flipping from beginning to ending and then to alternative endings with no concern whatsoever for the nonsequentialness of it all. "Choose Your Own Adventure" was an important book franchise.

Obviously, when they inserted "nonsequentially" into their contract, the folks at Putnam knew about hypertext, the software that enabled readers to jump around a text by clicking on hotlinks that could take them to unlimited alternate scenarios.

As if book-loving traditionalists weren't nervous enough, the nonlinear nature of hypertext freaked some of them out, and one began to hear mutters about the crumbling of the book industry and the death of paper. Robert Coover, in a *New York Times* article entitled "The End of Books," wrung his hands over the "dimensionless infinity" of hypertext software:

> The traditional narrative time line vanishes into a geographical landscape or exitless maze, with beginnings, middles and ends being no longer part of the immediate display. Instead: branching options, menus, link markers and mapped networks. There are no hierarchies in these topless (and bottomless) networks, as paragraphs, chapters and other conventional text divisions are replaced by evenly empowered and equally ephemeral window-sized blocks of text and graphics—soon to be supplemented with sound, animation and film. . . . If everything is middle, how do you know when you are done?[9]

Putnam's Display Rights was the first shot in what would eventually become a war, and I wondered when similarly aggressive language would begin showing up in other publishers' contracts. Would they seize the high ground as Putnam had done? As the '90s unfolded, I girded for combat.

But despite a number of skirmishes, war didn't break out, because there were not yet clear-cut sides, either buyers or sellers. The publishing community was not yet immersed in the e-book culture that would one day make uploading a file as much second nature as making a photocopy. The technological information was confusing and contradictory, and lacked comprehensible standards and protocols, leaving many executives paralyzed with indecision.

There did not yet exist the kind of shorthand nomenclature ("DRM," "RTF," "PDF," "JPEG" etc.) that welds workers together. There was no universal e-book reader, no handheld equivalent of a fax machine, no simple *something* that everyone could rally around and declare, "Ah, *now* I get it!" Even when

Stephen King, the towering commercial novelist of the age and a cultural icon, released his e-novella *Riding the Bullet* in 2000, billed as the first original mass market e-book, it was still considered by many to be a novelty; widespread acceptance of the medium did not follow.

Another factor impeding apotheosis of the e-book was resistance, if not downright hostility, to this new wonder. Many editors, writers and readers did not believe that e-books were books, and even long after the medium was demonstrated to be popular and profitable, distinguished authors (Stephen King notwithstanding) and their agents refused to permit their works to be converted to digital format, because it was *infra dignitatem*. The J. D. Salinger estate held out on e-book editions of *Catcher in the Rye, Franny and Zooey, Nine Stories*, and *Raise High the Roof Beam Carpenters* until August 2019.[10]

As I contemplated my options at the start of the 1990s, I realized that the e-book industry was far from established, and there were many paths yet to be blazed. But which ones? The fact was, I didn't know beans about computer technology. I couldn't write a word of code. The only thing I understood was books, or what the techies were now calling content. I recognized—a realization we would call "duh" today—that whatever avatar electronic books were to take, whatever the shape or form of the device, they would need books.

And I was sitting on a trove of them.

3. CONTENT AND DISCONTENT

(THE 1990s)

> But vultures in the air still hover,
> We throw out dough we can't recover.
> We genuflect to new Madonnas—
> Leonas, Marlas, and Ivanas.
> We buy and then in turn are bought,
> Ignoring what last decade taught.
> It looks as if this want of feck'll
> Dominate the fin de siècle.
> How quickly the suspicion grows:
> Plus ça change, c'est la même chose.

I SURVEYED THE HORIZON and speculated on where the content would come from to feed the digital beast slouching towards the publishing industry. The answer was obvious. I knew that readers would be hungry to acquaint, or reacquaint, themselves with popular genre authors whose out-of-print books could be found only by rummaging around secondhand bookshops. The prices charged for rare editions of beloved authors were often way out of range for all but professional collectors. Today a first edition of Harlan Ellison's 1993 Ace paperback *I Have No Mouth and I Must Scream* sells for about $185.00; the 1973 Ballantine paperback of *Tarnsman of Gor,* the first novel in John Norman's popular Gorean fantasy saga, costs almost $1,500.00; a new copy of the first edition of Dan Simmons's classic *Hyperion* will set you back almost $4,000.00. Even a book of my own, the novelization (under the pen name Curtis Richards) of John Carpenter's movie *Hallcween*, out of print for over four decades, goes for $1,395.00. The first edition of Ray Bradbury's *Fahrenheit 451* is priced at $45,000.00.*

Though I believed that critical mass in e-books would occur only when new titles were released, from a strategic point of view it made more sense to start with backlist (a trade term meaning previously published books). New and original books need a lot of capital to validate them, because there are no reviews or buzz to attract readers. With backlist books you don't have this problem. They are validated simply by virtue of the fact that brand-name publishers like HarperCollins or Simon & Schuster thought well enough of them to publish them. And for most backlist titles, reviews are available to promote them.

At that moment in time—the early and mid '90s—the only practical apparatus for reading electronic books was the desktop PC. Laptops had recently come on-stream but were heavy, expensive and technically limited. I reasoned, however, that genre fiction would appeal to fans so starved for their favorite authors that they could endure the tedium of reading them at their desks on computers. The term for these adventurous readers was "early adopters," and in the dawn of the e-book era they would drive the business until it caught on with conventional consumers.

I was aware that other literary agencies sat on stockpiles of dead or moribund books—a good many of them significant works by first-rate authors—and I lobbied my colleagues to review their contracts and get their rights back. I was editor of the newsletter of the Association of Authors' Representatives (AAR), an organization of several hundred agencies created in 1991 out of a merger of two earlier associations. (I subsequently became its president.) I published a number of articles and editorials heralding the New Age to my colleagues. I even formed a company in 1994, called The Content Company,† dedicated to handling e-book rights for agents who hadn't the time or inclination to do it themselves. If they were too busy to review their contracts, I wasn't.

Alas, all these endeavors came to naught.

The reasons for this failure were ignorance, apathy and inertia. Many agents and authors hadn't the slightest idea what I was talking about, or if they

---

* All price quotations from AbeBooks.

† The firm currently bearing that name is not related to the one I founded.

did, it was hard for them to grasp how any of this applied to their business and how they could make money on what was still an abstraction. I can't blame them. E-books in the early '90s must have sounded as speculative as bitcoins do to many lay people today. We have it on good authority that some investors are making fortunes in cryptocurrency, but the time and energy required to understand and profitably invest in it are way beyond the patience of most mortals (at least this mortal).

Take a hypothetical book, *My Life in Skinny Dipping* by Norbert Glottalstop. Let's say it was published in 1960 by a company named Coward McCann and your e-book outfit wants to reissue it. When you start inquiring, you discover that Coward McCann went out of business decades ago and it takes you two weeks to track down that it was acquired by G. P. Putnam, and another two weeks to learn that Putnam in turn was acquired by Penguin which became Penguin Putnam which then became Penguin Random House. It takes another three months for Penguin Random to locate the contract and three more to learn the whereabouts of Norbert Glottalstop. He's six feet under. You finally get hold of his widow who tells you in exquisite detail how poor Norbert drowned ten years ago when a giant clam seized his big toe.

"That's a darn shame, Ms. Glottalstop, but would you happen to know if your husband got a reversion of rights to his book?"

"Well, I never did pay much mind to Norbert's business dealings, but next time I go up into the attic . . ."

I could go on, but you get the picture. It's a nasty job, but someone's got to do it if you want that book back in print and you want to be sure there's a clear-cut chain of title. Multiply that story by hundreds or thousands, and you will understand how hard it was to motivate other agents.

Many agencies are old, thirty or forty years old or more. If not lost or misfiled, their innumerable contracts, amendments and correspondence are squashed into filing cabinets, the paper yellow and brittle, and photocopies almost illegibly faded. For a busy agent to pore over them one at a time, deciphering their suitability for this new medium, when no one was sure what this medium was or when if ever it would become relevant or profitable—life was too short!

To the above reasons you can add hostility. Many publishing people considered digital books to be a flash-in-the-pan novelty at the very least and a threat to their beloved Gutenberg mindset at worst. Some hated the idea so passionately that they refused to have anything to do with digital books, an animosity that endured well into the twenty-first century.

Bizarrely—for you would think they would welcome a potential new source of income and readership–this aversion extended to authors, for whom the only kind of book was a *book* book. Though I promoted the digital future in my monthly column, the idea didn't gain much traction.

Some authors complained, "Now that you got the rights to my books back, why aren't you exploiting them?"

I could only tell them, "Trust me. They're money in the bank."

Even science fiction writers dedicated to imagining future worlds had a hard time imagining the one being born in front of their eyes. My fantasist client Harlan Ellison, never one to mince words, brandished a copy of one of his hardcovers and flatly declared *"This* is a book. *That"*—jabbing his forefinger at my computer monitor—"is bullshit."

In short, it was just too early for any industry-wide updraft of enthusiasm. I had no choice, therefore, but to go it alone. So I set about painstakingly reviewing my contracts and royalty statements to decipher their status. This seemed like an easy enough task: if a book's royalty statements displayed zero sales for the last three or four semiannual reporting periods, it was out of print, right? All we had to do was ask the publisher to return the rights to us, right?

Wrong.

The contractual provisions defining "in print" in older books were rife with ambiguities, making it maddeningly difficult and time-consuming for publishers to give, or agents to get, confirmation that any given book was or was not "O.P." Requests for reversions thus went to the very back of the line, and one had to hound rights managers mercilessly to get answers.*

Here is a typical contract boilerplate definition of "in print":

> The book shall be considered in print if it is on sale by Publisher or under license granted by Publisher as provided herein, or if any contract for its publication is outstanding.

When you analyze it through the jewelers' loupe eye of an agent or lawyer, you will readily see that this wording raises more questions than it answers.

---

* Or romance them. I was more obsequious to rights and royalty managers than I was to CEOs.

What for example does "on sale" mean? Does it mean copies in bookstores? Copies in the warehouse? A listing in a catalog? How many copies? One hundred? One dozen? One? What if the only extant copies are just a handful moldering away in a warehouse for a decade? And how do you ascertain the number of copies in stock, anyway? Visit your publisher's warehouse?

What about that "license granted by Publisher"? Is a ten-year-old contract with a Turkish publisher sufficient for your publisher to justify hanging onto the rights to your book? Do you have to prove that there are no more copies in your Turkish publisher's warehouse? The nightmare potential was very high, and sometimes the nightmares came true.

I devised a three-part strategy for building an inventory of titles, and my tactic was elementary: naked and unremitting aggression.

The first task was to recover the rights of books that were undeniably out of print. I bombarded rights managers with letters and phone calls. I reasoned, coaxed, cajoled, whined, threatened, and generally made a pain in the ass of myself. I was gambling that publishing employees had not yet recognized the potential value of backlist books.

Perhaps they hadn't, but publishers nevertheless clung to their books like limpets. For one thing, they don't make money giving rights back to authors, so those requests fell to the bottom of the paperwork sea, where managers hoped they would eventually be covered up by the sediment of new paperwork and forgotten. More important, publishers reasoned that even if a book is out of print there is no telling when the author may get hot, at which point those dormant early books will have tremendous value. Thus, publishers deliberately created ambiguous out-of-print clauses that enabled them to hold onto the rights for as long as possible. Nevertheless, little by little my browbeating succeeded in dislodging the rights to several hundred books.

But there was also a hoard of titles that were still—at least arguably—in print. These constituted the second target of my offensive. I scrutinized every contract, hoping not to find those disqualifying phrases like "information storage and retrieval" or "now in existence or hereafter devised." Those that lacked them, I added to my storehouse. Those that had them required another strategy. I contacted the books' publishers and proposed licensing e-book rights from *them*, offering to split the royalties between themselves and the authors. In many cases they agreed—and provided us with cover art and jacket blurbs to boot.

The third front was new titles. In the hope that publishers were not yet hip to the potential value of e-books, I fought to withhold those rights when I negotiated deals for new books. Happily, some publishers conceded them to me, confessing (after some forceful prompting) that they were no more capable of producing e-books than they were of producing jet airliners. As one editor said to me, "I don't know what e-books are, but you seem to, so you can have them." I did, thank you.

And so, by hook or crook, I ended up warehousing over one thousand titles.

However, with stories about e-books showing up in book industry trade publications with more and more frequency, I knew it wouldn't be long before the window on publishers' acquiescence closed.

4. THE AGE OF MIRACLES

(THE 1990S)

Now that paper is de trop,
Digital is comme il faut.
Manuscripts and printed copy?
Deader than the three-inch floppy.
Editors eschew blue pencils.
Obsolete as quills and stencils!
To make our industry more green,
They edit on a Sony screen.
And agents now submit their schlock
By means of email as dot-doc.

S CIENCE FICTION MASTER Arthur C. Clarke famously said, "Any suffi- ciently advanced technology is indistinguishable from magic." I whole- heartedly concur. You can explain GPS to me till you're blue in the face, but you will never convince me it isn't magic.

The last decade of the twentieth century produced so many technological wonders on the road to e-books that no one could be faulted for believing they flowed from some mystical Fourth Dimension. In actuality they were an amal- gam of game-changing achievements conjured not by supernatural wizards but by scientific ones. Without any one of these developments the dream of e-books would not, could not, have come true.

I was one of those dreamers, but as I was involved in the quotidian tasks of running a business, I could scarcely afford to pay more than casual attention to the research and development in progress. And besides, my main interest was not the technology but rather the content itself.

Because of my inadequate scientific grounding I had only a general idea of what an e-reader would look and feel like, and in this, I was not alone. My own speculations were laughable: I imagined a kind of lightbox into which you inserted mini-DVDs, the same way you inserted compact cartridges or disks into Walkman audio devices, and you would then read them like a slide show: Click, next page. Click, next page. Click, next page. Years later, an early e-book device actually did click at each turn of the page. The consequences were not always agreeable. A colleague told me he was reading a book on the device in bed late one night while his wife slept soundly beside him. Click-click, click-click, click-click, click-click. After ten or fifteen of these she opened her eyes and said, "If you don't turn that fucking thing off, I will kill you." This design flaw was eventually rectified.

My ingenuousness at the time demonstrates that I was still mentally grounded in things tangible—hard disks, cassettes, desktop computers, etc. (At the time I was not aware of The Watchman, a 1½-pound portable television introduced into the United States by Sony in 1984. As it utilized cathode-ray tube display and an aerial to catch broadcasts, it never caught on and eventually succumbed to digital technology.)[1] The Web as the mother source of digital content was, in the early '90s, pretty nebulous. In this shortsightedness I don't think I was different from many of my book industry colleagues. But as the decade progressed, publishing people caught on and caught up, and by the end of it they were at least minimally tech-literate and ready to venture into the new world.

A thumbnail sketch of that world might help to put things into context.

Throughout the 1980s, the technology enabling devices to communicate with one another was called the Ethernet, and the connections were effectuated via cables or phone lines—hardwired conveyances as opposed to wireless. Modems converted digital information into coded signals. The signals were sent via cable and the information reconstituted by the modem of the receiving party. (MODEM is an acronym for "MOdulator-DEModulator"—modulator = encode and upload, demodulator = download and decode.)[2]

The Ethernet system was particularly useful in local area networks. These LANs served geographically close and homogeneous organizations like colleges or business offices. But though the Ethernet was reliable and secure, it was also slow and expensive, and its dependency on cable limited its reach. By the end of the 1990s the volume of information to be transmitted exceeded the bandwidth of the lines carrying it. The world looked for a speedier and farther-reaching means of intercommunication. They turned to wireless.[3]

Though it may seem as if it's been with us forever, it was only in 1993 that the Internet was born. In 1989 scientists Tim Berners-Lee and Robert Cailliau of CERN (a French acronym for the European Council for Nuclear Research) developed a system for applying hypertext to an information-sharing program. A year later they produced the world's first web server (their first project was a telephone book). They linked their system to the ARPANET, the network of linked computers created by the Pentagon in the 1960s. They named the offspring of these efforts the World Wide Web. In what Livinginternet.com called "a fateful decision," Berners-Lee persuaded CERN to release the Web from proprietary ownership, placing it in the public domain and thus available for public use in 1993.[4] Whereupon the Internet burst into bloom and literally transformed the world, enabling the spawning of legions of websites. In 1995, domain registration was instituted, and by 1998, registered domain names had soared to more than 2 million.[5] As of this writing there are more than 200 million active websites.[6]

As we have seen, hypertext, developed in the 1960s, is a software program that links texts to other texts, graphics, sound and video. (The formal name for the program is Hypertext Markup Language, or HTML.)[7] Before it was created, the protocols for access to a server were pretty primitive, "little more than glorified Word documents" in the words of one industry observer, Rease Kirchner, Content Marketing Manager of Webflow.[8] Using hyperlinks (also called hotlinks), a host of media can be displayed (or even performed) via devices ranging from stylus to touch screen to mouse. By the end of the decade, HTML had replaced plain text as the foundational format for building dynamic web pages.

In all likelihood, bloggers first popularized the tool's multifarious glories in their interlinking of text, photos, artworks, spoken word, sound effects, music, videos and animation on their websites. (According to Amanda Zantal-Wiener, the term "weblog"—soon shortened to "blog"—wasn't introduced until 1997, and "blogger" not until 1999. And it took till 2000 for videos to be embedded in blogs.)[9] Though some writers experimented with multimedia e-books, its use in authoring traditional fiction and nonfiction was more limited, and most writers continued to compose in plain text, using a variety of conventional word-processing software (and they still do).

Perhaps the biggest godsend of hypertext for authors was the ability to create websites. Until the last decade of the twentieth century, most authors were totally dependent on newspapers and magazines for reviews, advertising and self-promotion. Commercial clipping services scoured urban and rural publications for announcements, reviews, even mere mentions of authors and their books, and provided them—literally scissor-clipped and postal-mailed—to publishers or agents. But now, authors roving the Web could access these references themselves and produce their own personalized displays, featuring biographies, bibliographies, book jackets, reviews and photos. In just a few years, their careers and accomplishments became accessible to anyone who searched for them.

Balancing its benefits was the sprawling, inchoate, "dimensionless infinity" about which Robert Coover despaired in his *New York Times* article. He was not alone: five hundred years of plain text was not a habit broken overnight. Even authors were hostile to hypertext, preferring the purity of straightforward storytelling to what they decried as imagination-killing and narrative-interrupting links to illustrations and other distractions. In some quarters, pushback against the newfangled format was as fierce as resistance to talkies had been after decades of silent films.

But for a savvy young generation, exploring the potential of this new tool was like discovering new celestial wonders through infrared telescopes. Recognizing hypertext's significance, publishers began to apply it to a variety of functions such as website design, conference presentations and, eventually, electronic catalogs, e-galleys, manuscript editing software and contract processing. When e-books did arrive in the next decade, facility in markup code became a staple in the editorial toolbox. Though it would take several years for the book industry to fully realize its potential, hypertext transformed communications. For an analog-bound book industry, it was as if a fourth dimension somehow materialized before everyone's eyes.

It literally materialized before my own one morning in the late '90s, for as I was packing up a box for submission, a realization brought me up short. As I had done all my professional life, I placed the typescript into a custom-made box, added a sheaf of magazine and newspaper tear sheets with reviews and feature articles, laid a covering pitch letter on top, and prepared a manila envelope and label.

I stopped midtask and laughed. "What are you doing, fool?" This procedure was as antiquated as the Pony Express. Hadn't I just finished writing an article acclaiming the robust, kinetic, colorful new tools available on my computer? So? Why wasn't I submitting the book by email? Why wasn't I attaching the manuscript to the email (that problem had been solved in 1992)? Why wasn't I embedding hotlinks to biographical, bibliographical and promotional information, and videos of the author's appearances on talk shows?

I set the parcel aside and composed my first email presentation. I don't know if I was the first agent to do this, but I wasn't aware of anyone else doing it.

Thanks to the law of unintended—and sometimes comedic—consequences, editors did not quite know what to do with these content-laden emails. Though a growing number were versed in Internet communications, they still preferred to read manuscripts on paper. Should they print out the whole book and discover they hated it after one page? Should they print out a few pages only to discover the material was so compelling they had to get on the subway in the middle of the night, return to their office and print out the rest? (Don't laugh. It happened to me.)

A few editors contrived to read submissions on those newfangled Soft-Book tablets, but that had its drawbacks, too. "Almost all our editors live in Brooklyn," a publisher explained to me. "They like to read manuscripts on the subway from Manhattan. They can't do that on a laptop."

In time these contretemps were overcome. By the mid 2000s, submission of manuscripts via email and reading them on laptops were completely de rigueur—on the subway to Brooklyn or a plane to Paris.

And with every enhancement introduced, submissions became more colorful, dynamic, alluring and entertaining. Authors and agents created masterworks of artistic display with podcasts, video trailers, carousels and even buy buttons. At some point it crossed my mind that editors might be more dazzled by the bright shiny displays than by the quality of the texts. Surely, they were too savvy to acquire a book merely because the author was a hunk or a beauty, right?

In time, I was to learn the depressing answer to that question.

Hypertext and the Web were a match made in heaven. But several other elements were necessary to achieve critical mass for the e-book explosion. One was a means to enable users to locate specific websites on the Internet. As new web browsers were released, traffic on the World Wide Web swiftly expanded from only five hundred known web servers in 1993 to over ten thousand in 1994. That year a pair of developers, Jerry Yang and David Filo, created a Web

Directory to which, a year later, they added a search function, making Yahoo! the first truly popular search engine.

Many more search engines followed, among them an outfit called Rank-Dex, which, in 1996, was the first to use hyperlinks to index websites and gauge their quality. On RankDex's "About" page is this fascinating statement: "Many experts and commentators are unaware that RankDex was cited in Lawrence Page's patent application for Google PageRank as the first qualitative search engine."[10] PageRank was a program that employed an algorithm to weigh the comparative performances of various websites. Lawrence Page, better known as Larry, was to team up with his fellow Stanford PhD colleague Sergey Brin, in 1998, to launch the company that became Google, the proprietors of the most powerful search engine in the world.*

As the '90s progressed, the expansion of programs and applications challenged the existing means for storing them. Tape cassettes and CD-ROM optical discs were no longer adequate. A key solution materialized in the spring of 1997 with the introduction of the DVD disc, whose capacity was a giant leap over that of the CD-ROM. Whereas the earlier format stored about 700 megabytes of data on one disc, the DVD held as much as 17 gigabytes—that is, 17,000 megabytes, nearly 25 times the size of the old disc. (DVD originally stood for "digital video disc" but was changed to "digital versatile disc" because developers used it for other applications besides video, notably, music and document storage.)[11]

Unfortunately, even a 25-fold increase in storage capacity was dwarfed by the hunger for space created by new media. Storage needs for video games, for instance, soared into double-digit gigabytes and eventually into the hundreds. As storage needs grew, the inconvenience of swapping out disc after disc to play games or movies became a growing irritant for users. The limitations of a PC's RAM (random access memory) ruled out the home computer as a feasible repository.

It was now clear that the ideal—the only—purveyor of content had to be the Internet, and by the middle of the 1990s e-commerce was beginning to boom. Video games generated some $30 billion in 1998; eBay and Amazon, soon

---

* Google also boasts the rare distinction of being a proper noun (with a capital G) converted to a verb (lowercase g) to describe the act of browsing the Web, joining the company of popular generic products like Kleenex, Xerox, Frigidaire and Band-Aid.

to be dominant e-commerce stores, debuted in 1994 and '95 respectively, and in 1998, PayPal was launched as a means of processing online transactions.[12]

The flood of technological innovations set off tumultuous conglomeratization of the media, shrinking a pool of hundreds of companies to a mere ten at the start of 1990.[13] Three years later, the advent of the Internet unleashed a new media feeding frenzy.

Alarmed by a growing Wild West anarchy in the regulation of media, especially the Internet, Congress passed and President Bill Clinton signed the Telecommunications Act of 1996, legislation aimed at overhauling regulation of the media.

The high-minded purpose of the act was to increase competition, open the communications arena to new businesses, deregulate cable fees, and end the prohibition of television networks owning cable systems. The act's implementation, however, had the opposite effect. "By means of deregulation," Michelle Kratz wrote in a cogent thesis for Pace University's Dyson College of Arts & Sciences, "the act was meant to increase market competition and decrease prices for consumers. In reality, the number of media subsidiaries (particularly television stations) one company could own was raised significantly, thereby prompting many of the larger media companies to purchase substantial portions of the market."[14]

Among those "substantial portions" were American publishers constituting four of today's Big Five:—Time Warner (Little, Brown and Warner Paperbacks, currently Hachette); News Corporation (HarperCollins); CBS/Viacom (Simon & Schuster); and Bertelsmann (Penguin Random House). The exception was Macmillan, a wholly owned subsidiary of the multibillion-dollar Holtzbrinck Publishing Group. As we shall see, these contractions of the once sprawling publishing community would have colossal implications for the e-book business.

Inevitably, publishers turned their attention to the Internet as a means of marketing conventional books. They were about to get far more than they asked for.

Entrepreneur Jeff Bezos surveyed the book business and beheld the very essence of an analog industry. Its products were tangible artifacts distributed in fossil-fueled vehicles to brick-and-mortar stores. Over centuries, countless numbers of books had been published, yet only a fraction was cataloged, and only a fraction of that was available to the public. Though a great library or superstore might carry tens of thousands of unique titles, the challenge of locating them was arduous and time-consuming. And 10,000 or even 150,000 titles (the capacity of some book superstores) were drops in the bucket compared to the tens of millions estimated to have been published.

Bezos's vision was an online marketplace with an unlimited catalog that a customer could quickly search, select, order and buy, eliminating the necessity of a visit to a store or library. Required was a robust website, employing the quartet of components we have described: hypertext, the Internet, information storage and search engine. They all came together in amazon.com, which he launched in July 1994.

To those ingredients he added marketing, enhancing not just publishers' and consumers' experience but authors' as well. The term "discovery" was applied to describe the fortuitous encounter of books and authors through online browsing. And Bezos saw profitability in the efficient locating and purveying of hard-to-find books.[15]

Before long, major frontlist publishers cast their lot with Amazon, which not only sold their books but provided publication data, critics' quotes, and customer reviews, and ratings and rankings, making it the go-to source for book information. He added used books to the inventory, enabling third-party booksellers to offer their wares on the Amazon website in exchange for a commission. The acquisition in 2002 of AbeBooks, one of the largest used-book marts in the world, dramatically extended Amazon's control of the marketplace.

In light of the "Everything Store" that Amazon has become, and its reputation for disruptive and rapacious business practices (about which, more in due course), it is difficult to recall that the company was once a small start-up focused on nothing but physical books. (Amazon did not make its first nonbook-related acquisition until 1998.) When Bezos opened his website for business, he declared it to be "Earth's biggest bookstore."[16] In truth, it was a pygmy compared to Barnes & Noble. By the end of 1996, its first full year of operation, Amazon reported revenues of $15.7 million, compared to B&N's 2.448 *billion*.[17]

Amazon's significance in the final years of the twentieth century, therefore, was not as the colossus we know now but as a trailblazing model of Internet commerce, one that fulfilled the dreams of the founding fathers and mothers of the World Wide Web.

However . . .

To whatever extent Amazon revolutionized bookselling, the product it carried was nothing other than good old book books, bringing us no closer to e-books than we were in 1992 or, for that matter, 1492. Yes, you could store them on disks, and unlike other forms of entertainment, such as music and movies, books did not take up a lot of digital storage space. A single DVD could carry over 50,000 of them, far more than most mortals could consume in a lifetime. But for all their capacity, DVDs were still tangible objects designed to be inserted into machines.

Nor was there yet a satisfying way to read them. Desktop computers were out of the question for all but the most dedicated booklovers. Laptops were a little better but still nothing you wanted to curl up with in bed. Desperately needed was a means to untether users from their desks. It kept coming back to handheld e-readers.

In 1998, we got them.

5. TURNING POINT

(1998)

The trade enjoyed increasing traffic
In comics, manga and novels graphic
As grownups hastened to embrace
The dumbdown of the populace.

Y OU WOULD THINK THAT the obvious choice for a handheld e-reader
would be the cell phone. The device had steadily grown more powerful and
versatile since the 2.5 pounder with a twenty-minute battery life introduced in
the 1970s. Of critical importance was the development of the lithium ion battery
in 1991, which was to become standard in most cell phones and a giant step
towards miniaturization. These appliances had shrunk from the one used by
Gordon Gekko in the 1987 film *Wall Street* which, as Isabelle Raphael described
it in *Parade*, "was nicknamed 'The Brick.' . . . Given its huge price tag [nearly
$14,000 in 2025 dollars]," Raphael added, "it is not surprising that only Wall
Street Masters of the Universe like Gekko could afford it."

The term "smartphone" was not coined until 1997.[1] However, the IQ of
cell phones soared from bright to off the charts in the '90s, adding touchscreens,
keyboards, calendars, cameras, address books, calculators, notepads, gaming,
emailing, texting and other features, all packed into a device not much bigger
than a deck of playing cards.[2] However, even after 1996, when a Nokia cell
phone first connected to the Internet, phones proved inadequate for accessing,
storing and displaying dozens let alone hundreds or thousands of books.[3]

A significant exception was growing experimentation by the Japanese with manga on cell phones. This genre of pictorial storytelling literature dated as far back as the twelfth century. In the 1980s and '90s technologists began to apply computer graphics to this hand-drawn art form. The natural evolution, which took place in the first decade of the new century, was the cell phone novel, narrated in short installments of no more than 160 characters, about 50–100 words. The novelty became a rage in Japan, and some of these works became bestselling print books, hit movies and even television series. In 2008, the phenomenon reached America's shores in the form of a website called Textnovel.[4]

But cell phone novels and Web manga were not to fulfill the dreams of e-book developers. Throughout the 1990s, I scanned the book trade publications for announcements of strides in e-reader development but found little hard news. I thought of inventing one myself, an undertaking that would surely have driven my family into abject poverty, but I abandoned that scheme when Leslie threatened to banish me to another continent. We now know that a number of entrepreneurs were developing handheld e-readers, but their efforts were shrouded in secrecy, obviously out of the well-justified fear that competitors would beat them to the pot of gold. Among them were Jeff Bezos and Steve Jobs. However, it would be years—2007 and 2010, respectively—before their killer creations would be rolled out.

But several enterprising technologists announced that they'd found a path to the prize and, within months of each other, introduced their products.

One was Peanut Press—the answer to a riddle posed by Matt Richtel of *The New York Times*: "What do you get when you cross the printing press with the Palm Pilot?" That gadget, first released in 1996, was one of a variety of organizers known as PDAs (personal digital assistants) devoted to managing—in this case with a stylus—such information as contact lists, calendars, recipes, documents and financial spreadsheets. Mark Reichelt and Jeff Strobel, Peanut Press's founders, devised a way to import digital books into the Palm Pilot via desktop computer. Unlike Project Gutenberg, which carried only public domain titles, this e-book publisher acquired rights from major publishers to copyrighted books that were still in print. Though their screen was small ("the width of a magazine column," Reichelt said) and the device not fed by the Internet, at 5.7 ounces, the original Palm qualified as a handheld. It ran on two triple-A batteries.

In 1998, two other candidates, in look and feel superior to anything to date, made their debut.

The first was SoftBook. At 11×8 inches, with a 9.5-inch-diagonal touch-screen, the tablet-size SoftBook was never intended to be a handheld unless you held it in both hands and rested it on your stomach: It weighed in at 2.9 pounds (including its handsome leather case). The size made it suitable for magazines and newspapers as well as books, and it carried *The New York Times, Bloomberg News, Time* and *Newsweek* among other papers and periodicals. The text was black-and-white but not grayscale.[5] A subsequent version, the REB 1200, had a color screen.[6] The price for the original version was around $1,200 in today's dollars. It held about 1,500 pages but could be expanded with a flash card to somewhere between 50,000 and 85,000 pages. The user did not need to link to a desktop computer to access books and magazines but instead downloaded them from the vendor via a phone jack—a process that sometimes took over-night. You would then have about five hours to read before having to recharge the battery. Despite these shortcomings, a number of major publishers signed up to license their content to SoftBook.[7]

The second new candidate was NuvoMedia's entry, the Rocket eBook. It looked like a husky prototype of the Kindle and had a screen 4.5×3 inches, but was too bulky to fit into one's pocket, weighing 1.25 pounds (*excluding* its hand-some leather case). Priced like the SoftBook (nearly $1,200 in today's dollars), it could hold about ten books and gave you plenty of time—up to forty hours of battery life with the LCD backlight off or twenty with it on—to read them using page-turning buttons. It had sexy stuff like text-search capabilities, bookmark-ing and highlighting, and a virtual keyboard you could use to make notes (using either your stylus or your finger). Robert Carter, Senior Vice President and head of content and business development, tells me he rounded up some three thou-sand books from HarperCollins, Simon & Schuster, Random House, Macmillan, Doubleday and numerous other publishers including my own company, E-Reads.

Originally, according to Carter, the name of the gadget was simply Rocket Book, but the owners were prevailed upon by Jack Romanos, President (and soon to be CEO) of Simon & Schuster, to change it, because it sounded like Pocket Books, S&S's mass market paperback division. So NuvoMedia added the *e*.

The Rocket eBook was rolled out with great fanfare by Barnes & Noble. The bookseller and Bertelsmann (current owners of Penguin Random House) had each invested two million dollars. In its first year, some twenty thousand units were sold. The professional version introduced in 2000 had a capacity of about three dozen books.

Brad Stone, in his book about Amazon, *The Everything Store*, tells an intriguing corporate story behind the Rocket eBook. Creators Martin Eberhard and Mark Tarpenning first pitched their prototype to Jeff Bezos, but he rejected it because he didn't like that it had to be attached to a computer in order to upload books. Eberhard and Tarpenning went on to found Tesla Motors (long before Elon Musk was a gleam in Donald Trump's eye).[8] Subsequently Bezos had a second opportunity to invest, but because it meant supporting bookselling rival Barnes & Noble, he passed again. But he was obviously intrigued by the device and at some point set out to develop his own, a gadget that surmounted all the shortcomings of its predecessors.[9]

He succeeded—nine years later.

One more product bears mentioning: the EveryBook Dedicated Reader with a full-color screen was announced for 1999 release. It never made it.[10]

The dawn of e-readers can arguably be dated 1998, even though, for a number of reasons, they turned out to be, as the carnival barkers used to say, "close but no cigar." Some were tethered to desktop computers, some too heavy or boxy or slow or small-screened or dim-screened, or they had limited battery life, poor navigation or complicated command functions. All in all, they were not quite what the world meant when it visualized "e-books."

But there was a far deeper issue: The world simply wasn't ready to embrace the precipitous transformation from printed books. Few titles had been digitized, and few consumers—even early adopters—were ready to take the plunge with an expensive item that could not provide a glitch-free experience. Only two years after they launched, both SoftBook and NuvoMedia unloaded their businesses to Gemstar-TV Guide International, which, not long thereafter, discontinued production altogether. I rescued one of each device to donate to some future e-book museum, along with the Fossil Rocket eBook watch handed out by Barnes & Noble as a launch-party favor. Unlike the e-books themselves, the watch still works (it is on my wrist as I write this).

Lots of other companies were working on the challenge, and the prospects were good as long as venture capital flowed, interest rates remained low and tax incentives fed the entrepreneurial appetite. A few Cassandras observed that tech stocks seemed to be inflating dangerously on the wings of speculation, and if the bubble burst, a lot of undercapitalized companies could find themselves overextended.

But this was 1998, and the future for e-books looked rosy.

In a perceptive July 1998 *Fortune* magazine article, Carol Vinzant summed up the shaky state of the art:

> The introduction of these products, no matter how successful, is hardly the simple, happy ending to the electronic-book story. As with most good tales, there will be sequels. The new e-books will inevitably look dowdy and expensive after better, thinner displays appear. . . . Electronic books will truly catch on when their technology and their complicated pricing structures become invisible to consumers. "The thing I really await," says Project Gutenberg founder Michael Hart, "is the paperback-sized text reader you can buy at Kmart for $20, with 200 books already loaded in."

"You may have to wait a while for the blue-light special," Vinzant concluded, "but the e-book, finally, has left the shelf of science fiction."

Taking all things together, by the turn of the twenty-first century, the essential reading implement was still the printed book. But the cascade of cell phones and smartphones, e-readers, personal organizers, game consoles and the like had profoundly impacted our culture in a very specific way: The act of navigating content on handheld devices had become commonplace.

Verily, these immense strides qualified 1998 as electronic publishing's *annus mirabilis*. The world was ready for e-readers. So was I.

6. CONSOLIDATION

(1990–99)

Huge conglomerates expanding
Till scarcely anyone's left standing.
Is it possible we're heading
Toward one great climactic wedding,
When all but two remain unmerged,
The rest absorbed, acquired, or purged?
The final stage of evolution,
The ultimate event of fusion,
A blinding flash, a cosmic bang,
The Yin becomes one with the Yang.

A S THE WORLD HURTLED toward a digital future at the end of the twentieth century, the traditional book industry lumbered down its own path, its denizens contemplating the imminent transformation of their business with a mixture of excitement and apprehension.

Encumbered by five hundred years of tradition, they had plenty to be apprehensive about. The very culture of the business now seemed to be fulfilling the disturbing scenario Thomas Whiteside had portrayed in his 1981 book *The Blockbuster Complex*, a searing analysis of publishing's transformation to a conglomerate mentality. Not for love or money could you come up with a better catchphrase than Whiteside's title to describe the prevailing mindset. The bottom-line character of the modern business stood in stark juxtaposition to the image of publishers as handmaidens to the muses, an ideal that every young editor entering the ranks still cherishes.[1]

The decades leading up to Whiteside's book were far from stable. More than three hundred publishing company mergers had been forged in the previous twenty years, according to an October 1977 *New York Times* article by Ann Crittenden with the provocative title, "Merger Fever in Publishing." Some of these consolidations even attracted the antitrust scrutiny of the Department of Justice.[2] Crittenden writes:

> Seven paperback publishers now control the bulk of the mass paperback industry, and all of them are part of larger corporations. Ten companies accounted for 89 percent of all book-club sales in 1976. According to industry analysts, only 40 of the estimated 6,000 hardcover trade houses in the country can successfully publish a book on a nationwide basis. Moreover, even these houses are rapidly being absorbed by larger conglomerates, such as Gulf + Western, or are themselves becoming the nucleus of a conglomerate. Doubleday, for example, now owns a paperback house and a book club.

Though the 1970s publishing business that Whiteside portrayed was turbulent, it was low on the Richter scale compared to the convulsions of the next two decades. Dozens of struggling presses merged with one another to consolidate their buying power. But even they succumbed to the malady of underfinance, leaving the industry's middens littered with combines like Lothrop Lee Shepard and Coward McCann & Geoghegan. Mergers and acquisitions were spearheaded by conglomerates that gobbled up trade book houses for their glamor, then spat them out for failing to maintain the 15 percent annual profit earned by other companies in their portfolios, the ones that made breakfast cereal and laundry soap.

At the end of the 1980s, wealthy foreign publishers swooped in to buy U.S. houses, and leading imprints became the possessions of the Japanese, French, Germans and Australians.

Of the Big Five today, Hachette is owned by the French, Penguin Random House and Macmillan by Germans, and HarperCollins by Australians. Only Simon & Schuster, acquired in 2023 by Kohlberg Kravis Roberts & Co., is owned by Americans. Starting with hundreds of significant publishers at the beginning of the 1980s, only two dozen or so were left at the end of that decade, and six by 2010.

Now we are five.[3]

On the positive side, a number of innovations had improved editorial conditions since Whiteside depicted the industry as "riddled with inefficiency, often sluggish management, agonizingly slow editorial and printing processes, creaky and ill-coordinated systems of book distribution and sales, skimpy advertising budgets, and . . . an inadequate system of financing." A host of software tools expedited almost every process from website development to cover design to catalog production. Desktop publishing enabled authors and editors to produce typography and typesetting on their personal computers.

The quality of these features was comparable to commercial printing but performed with a fraction of the time, labor and expense required back in the day. Web search engines stimulated the creation of social media platforms that brought editors, authors and readers together. And print on demand, first invented in 1990, was about to become commercially viable, solving a problem that had plagued publishers for centuries: how to maintain a book's inventory when its initial printing is exhausted. Simon & Schuster's Jack Romanos was one of the new executives perspicacious enough to mobilize the troops for the coming change. "The impact of technology on the prepublication process provided the publication with the files necessary to create e-books," he told me. "This allowed for the stockpiling of tens of thousands of new releases while we waited for effective e-book readers to be developed. Strategic investment in digitizing our backlist and develop effective digital rights management systems positioned most publishers for the coming revolution. I remember having to convince CBS to give me six or seven million dollars on a promise that there was going to be a future for e-books."[4]

Yet publishers agonized over the unknown consequences of the emerging world. And for good reason: The consequences were unknowable, a condition detrimental to the peace of mind of businesspeople. Left to their own devices, publishers would have been content to evolve at their customarily measured pace and gradually adjust to the changes coming out of the tech sector. But those developments were coming so thick and fast, in so confusing and contradictory a way that it was impossible for management to choose a path that would land their company on fiscal *terra firma*.

No business model existed that publishers could turn to for guidance, and few executives had a clue as to how, or even whether, e-books would fit into their business culture. How would they be produced, and who would produce them? How and where would they be displayed? What effect would they have on traditional print formats? What were they worth? Who was the audience? Where would e-books be sold? What royalty would be paid to authors? Would the digital edition be published before the print version or after? At the same time? Instead of?

In the absence of clear answers, publishers struggled to formulate policies and fashion them into cogent contractual language. The highest priority was gaining control of the rights, and they began erecting defenses along the lines of Putnam's Display Rights provisions, broadly encompassing every contingency they could think of to maintain hegemony. They drafted language according to themselves control of digital rights as well as a favorable royalty split.

The window of opportunity for authors and agents to recover rights to out-of-print books started closing and would presently slam resoundingly shut as publishers awakened to the treasures inherent in their backlists. Reversion requests were ignored, protracted or refused altogether. My own efforts in that theatre of war degenerated into hand-to-hand combat. If some sort of organized resistance was not mounted soon, authors would soon find themselves at a permanent disadvantage.

Agents had been slow to grasp the implications of the growing paradigm shift. With all the ink spilled about e-books in the trade magazines, if you didn't know something huge was coming down you had to be living in a cave. The issues were too big to ignore and were now topics of growing urgency.

Despite my importuning, the agents were fatally dilatory about putting up concerted resistance. With no business model, no financial structure, no sales data and no performance projections, there was no way for them to put numbers into the blanks when bargaining for fair compensation. As we have seen, with language and terms all over the place, the old contracts were no help at all. Everybody wanted to draw a line, but the only place to draw it was in ever-shifting sand.

Even if they had better information, the agents were seriously handicapped. Inasmuch as the AAR was not a guild, it could not undertake collective bargaining. The organization's cautious legal counsel instructed members to be careful not to run athwart of price-fixing laws. It was permissible for agents to *individually* negotiate e-book royalties, but to do so *collectively* risked government sanctions for restraint of trade.[5]

6. CONSOLIDATION

I strongly believed that author compensation for e-books should be identical to the division of book club revenue, namely a 50–50 split. Unfortunately, not only was I discouraged from explicitly promoting that percentage to AAR's members, but the organization's conservative policy forbade me from even advocating it at meetings. I murmured my subversive recommendations to members just out of earshot of the recording secretary, but the militancy I sought never materialized.

By the time the clouds parted, my worst fears were realized: the publishers had stolen a march on authors and agents, dealing themselves control of e-book rights and a lion's share—75 percent—of the format's revenue. As of this writing that is the so-called "standard" division of the e-book pie, and it's now baked into the boilerplate of every trade book publishing contract I know. Forgive me for wondering aloud how all publishers could arrive at exactly the same percentage when fearful agents dared not utter a target royalty aloud.

We did, however, win one very significant battle: We managed to reform the archaic termination language in book contracts. As we've seen, under the old regime, the vague and ambiguous language of out-of-print provisions enabled publishers to hold onto a book indefinitely. But the advent of digital technology offered an opportunity to precisely define a book's in-print or out-of-print status. Under the new definition, if sales of the print or e-book edition (or a combination of both formats) fell below a certain threshold, the author or agent could request a reversion of rights. The threshold was negotiable—it could be a dollar amount or a number of copies, but whichever it was, at least it was a discrete number that could not be disputed.

This triumph was overshadowed by far greater events, for the traditional trade book business was in the throes of vast upheaval. A trio of business storms caught the industry in a cyclonic updraft, carrying it far afield from the courtly profession of an earlier (if not mythical) era.

The first of these traumas was the consolidation of the handful of powerhouse publishers that remained standing. Mergers and acquisitions had taken a heavy toll on small and medium houses since the 1970s. But after passage of the Telecommunications Act of 1996, buying and selling of publishing corporations accelerated like *The Blockbuster Complex* on speed, sucking up many extant small fry that could not afford the resources to compete with a handful of publishing goliaths owned by deep-pocketed parent corporations.

But now the goliaths themselves were in play. Every one of what as of this writing (at least, as of this morning) is today's Big Five was bought, sold, merged or reinvented in the 1990s. The action was dazzling and far too complex to chronicle here. But to summarize:

- In 1990, News Corp, which had bought the British house William Collins, incorporated it with Harper & Row to form HarperCollins.
- In 1995, German publishing giant Holtzbrinck acquired the Macmillan Group, which today includes St. Martin's Press, Tor/Forge, Farrar, Straus & Giroux, Henry Holt and Flatiron.
- In 1996, the various acquirees and mergees in the Time Warner galaxy were brought together to form Time Warner Trade Publishing, which was acquired by online service provider AOL at the end of the decade and eventually sold to the parent company of Hachette.
- In 1998, Bertelsmann, which had earlier acquired Bantam, Doubleday and Dell, announced its intention to acquire Random House from Advance Publications.
- In that same year, Simon & Schuster, *née* Gulf + Western Inc./*née* Paramount Communications, was acquired by Viacom Inc.[6]

Whether you were buyer or seller, the disruptions were deeply disturbing, upsetting stable systems and supply chains, and putting many people out of work. The first victims of mergers were redundant services like printing and distribution, marketing, publicity, legal and accounting. Although it took longer for the Turk (as the deliverer of pink slips was sardonically nicknamed) to find his way to editorial offices, once the brass recognized that it made no sense to have two or three romance or science fiction or mystery imprints under the same roof (which was after all the purpose of merging to begin with), down came the scimitar.

The closing of imprints and discharging of editors resulted not only in the reduction of the number of published books but of the authors who wrote them. "When publishing houses merge," consultant and business analyst Sophia Jones was quoted on the publishing site Book Riot, "the first thing that happens is that the companies will combine their lists. This mean that they will take a look at their combined catalogs and decide which titles to keep, which titles to drop, and which titles to add to their respective lists. These decisions are made based on factors like popularity and profitability."

Jones added: "With fewer publishing houses, there will be less competition for authors and books, which means that the remaining companies have more power to set prices." Authors "may also find it harder to get their books published at all."[7]

That thought echoed years later when, in 2022, the Department of Justice stated, in a formal complaint against the proposed merger of Penguin Random House and Simon & Schuster, that it "would eliminate this important competition, resulting in lower advances for authors and ultimately fewer books and less variety for consumers." DOJ's position was upheld by the court, and the merger was terminated.

A second source of turmoil in the 1990s was the proliferation of bookstore chains and superstores. By far the most prominent actor was Barnes & Noble, whose expansion was nothing short of explosive. In 1987, B&N, already a behemoth, acquired B. Dalton Bookseller (798 outlets) and, in 1990, Doubleday (90). By the end of the decade B&N boasted 1,099 stores, and competitors Borders and Books-A-Million added hundreds more, saturating cities and suburbs.[8]

It was not just the sheer numbers of stores but their mammoth size: The average superstore was six times as big as a traditional bookshop and carried as many as 150,000 titles. These marts were conveniently located, attractive, well organized and staffed with informed clerks rendering knowledgeable and efficient service. Amenities such as coffee bars, lecture rooms, children's corners and even lounge chairs for browsers were provided to make these bazaars irresistible destinations. They offered something else, too: discounts, and if the lounge chairs and cherry danishes failed to lure customers away from Ye Friendlie Local Bookshoppe, a 10, 25 or 40 percent price cut certainly did.

Given how decisively Amazon has come to dominate book retailing, it's natural to think of bookstore chains like Barnes & Noble as underdogs striving against the odds for a place in the sun. But at its zenith B&N ruthlessly held publishers in a hammerlock to which countless small houses succumbed, and their rapacious policies contributed to the blockbuster mentality. For instance, in order to get an advantageous position in a bookstore for displaying an important book, such as a table at the front of the store, a publisher had to pay a high and sometimes exorbitant "co-op advertising" fee. "Pay-for-display programs are nothing new in the retail world," wrote Randy Kennedy in *The New York Times* in June 2005. "Supermarkets have long extracted money from manufacturers to put their boxes of cereal or detergent in eye-catching spots."[9]

But the practice seems less savory in bookselling, where bookstore
owners and managers were once assumed to serve as an editorial

presence, recommending and featuring books they liked. Besides, publishers complain that, despite its name, cooperative advertising is not a cooperative exercise in the least. Some compare it to a tax or even to extortion—evoking the practice of "payola" in the radio industry. Which is not to say that co-op is actually under-the-table, illegal or even unethical—it's just that bookstores don't tell customers about it.

"Publishers complain bitterly," Kennedy concludes, "that display programs are just another way that the big bookstores are dictating how they do business."

Indeed, the chains wielded so much power that they were able to influence or even control editorial decisions. Publishing sales representatives called on chain buyers as a matter of course to solicit support for acquisitions and marketing commitments. I have never forgotten my dismay the first time I was informed that a book I had submitted to a publisher first had to be presented to Barnes & Noble for an opinion before a decision to buy was made. In time, I learned that the store rendered judgment not only on submissions but also on cover art and even book titles. Barnes & Noble virtually sat on the boards of major publishing companies, and their thumbs-up or thumbs-down determined a book's fate as decisively as any editor-in-chief's.

Thanks to the wildfire expansion of bookstore chains, total bookstore sales rose over 55 percent, from $8.33 billion in 1992 to $14.88 billion in 2000. However, beneath the dazzling bottom lines were some symptoms of growing stress. The most telling was a dramatic rise in returned books.

As I explained earlier, thanks to the goodwill of publishers long ago, bookshops have customarily been able to return unsold stock to publishers for full or close to full credit, a practice unheard of in most retail businesses. However, even though returns have been a perpetual pain in the bottom line for publishers, it can generally be said that during most of the twentieth century a variety of conditions, such as smart selection of titles and prudent management of print runs and inventory, held returns to a manageable level. But all that conservatism got trampled in the go-go years of the 1990s, when both publishers and chains, seized by some inexplicable fit of mass optimism, lost control of fiscal restraint.

Publishers contributed to this *folie à deux* by throwing obscene and often unrecoverable advances at star authors, celebrities and flash-in-the-pan media phenoms represented by heavy-hitting agents. They hyped huge printings and offered (or at least accepted) easy settlement terms to induce chain store buyers to commit to big buy-ins. Once the chains made that commitment, the publishers proceeded to overprint.

Confusion was worse confounded by publishers grossly exaggerating the size of printings. Edward Nowatka, writing in *Publishing Perspectives*, pointed out that publishers announce "wildly inflated first printings in order to generate marketing buzz and buy-in from the booksellers. In reality, fearing returns, publishers often underprint from the announced number in order to hedge their bets."

The industry's worst-kept secret, these bloated projections were promulgated with cynical winks by trade news publications as "announced" printings as opposed to actual ones. "The fact is," Nowatka concluded, "that it is simply impossible to believe publishers' announced printings."[10] Editor and publishing consultant Stephen S. Power suggested announced printings were three or four times the size of the actual ones.[11]

Bookstores did little to restrain publishers from their profligacy. With miles of shelf space to fill in hundreds and hundreds of newly erected superstores, the chains often overordered highly touted titles with no concern about returning them, because—what the hell—there was no penalty for doing so. Publishers were choking on their own returns, some reaching as high as 60 percent. Doreen Carvajal, writing in *The New York Times* in August 1996, cited some disturbing statistics:

> For some titles, discarded books are spewing back to publishers at rates as high as 40 percent of gross sales. . . . Over the last five years, print runs for bestseller candidates have been inflated from an average minimum of 50,000 copies to at least 100,000, and the shelf life of some other books has been compressed from four months to as little as four weeks.

Most striking of all was this fact: "From 1990 to 1995," Carvajal wrote, "the industry's losses for rejected hardback books increased by 60 percent, to $531 million, while gross sales increased by 47 percent, to $1.64 billion."[12]

Publisher Alfred Knopf's mordant epigram said it all: "Gone today, here tomorrow."

Unable to stand up against the relentless march of the chains, independent bookshops went out of business left and right. Besides the chains' financial advantages, they co-opted the indie stores' reason for being, for many were specialty shops: feminist, Black or LBGTQ. "With B&N willing to carry these formerly taboo books, mainstreaming these subcultures, there was increasingly no reason for these stores," says Power.[13] According to the Book Industry Study Group, the combined market share of independent bookshops dropped from 44 to 19 percent between 1983 and 1994.

Complaints were being voiced that big publishers were giving preferential discounts and other special treatment to the chains. In 1994, the American Bookseller Association sued several publishers. The following year, three of the five publishers settled, and though no wrongdoing was admitted, they agreed to change their policies "to ensure equal discounts and promotional money to chains and independents alike," according to Patricia Holt in an sfgate.com article.[14]

An even bigger storm broke in March 1998 when the ABA and 23 independent bookstores filed a suit against Barnes & Noble and Borders alleging monopolistic practices, such as requesting special discounts from publishers. "The two companies," said Avin Mark Domnitz, Executive Director of the American Booksellers Association, 'are moving toward monopoly status: as stated in the lawsuit, Barnes & Noble increased its number of superstores from 135 to 469 in the last four years, while Borders jumped from 31 to 189. Of a $23 billion dollar industry, these two chains alone make up $5 billion."

Not surprisingly, the chains fought back, asserting that their huge revenues testified how well they were serving their customers. It took several years for the case to make its way through the courts, but ultimately, it too was settled with no prejudice on either side.[15]

While all this disorder was roiling, yet another volcanic upheaval was underway, one that impacted readers, writers and publishers alike, and was in some ways the most calamitous. This was the collapse of the market for paperback originals, works of genre fiction written specifically for the paperback market in such genres as mystery, science fiction, westerns, horror and women's fiction.

Introduced in 1939, "pocketbooks" took hold in the '50s and '60s, but publishers soon realized that the existing hardcover distribution system didn't work. They needed a different sales model, and they turned to the one used for magazines. Every month, magazines were shipped to depots—"agencies"—

around the country. Drivers picked up their quota of magazines from the agencies and visited stores on pre-assigned routes in towns and cities. Most of the stores were not bookshops but rather grocery, drug and candy stores, newspaper stands, and bus or train stations—wherever magazines were sold. Each month, these drivers removed the previous month's unsold publications from the shop's wire racks and replaced them with the current month's stock. (This explains why mass market paperbacks are released on a monthly basis as opposed to the seasonal basis for trade books.) Because the drivers in the mid '50s opted out of salaries in favor of commissions, they were generically known as independent distributors—"IDs"—and colloquially, as "rack jobbers."

To paperback houses, this distribution network was the perfect vehicle for delivering their product to a far-flung readership. An additional benefit was that often the drivers knew the reading tastes of all the accounts in their territory and stocked their vehicles with titles targeted very specifically at the audiences along their routes, even down to the names of the folks who watched for the latest book by their favorite author to go up on the racks. Some fans actually waited outside the store to intercept drivers as they made their deliveries.

A number of leading authors, sponsored by their publishers, were invited to visit the wholesale agencies and pitch their books to executives and jobbers. Some of the more enterprising authors went so far as to drop in on drivers early in the morning as they loaded books into their vehicles, bringing coffee and donuts and promotional material to inspire them. This technique was particularly successful with romance fiction. It did not hurt if the authors were attractive. Many a lovestruck driver stocked extra copies of a romance after its pretty author shared a predawn breakfast with him on the tailgate of his station wagon. It was a hallmark of superstar Jacqueline Susann's success.[16]

Ironically, it was their very prosperity that doomed the culture of mass market originals.

At the outset of the 1990s the paperback publishers were thriving. Rich paperback houses like Bantam, Dell, Warner and Pocket Books paid millions for reprint rights to hardcover bestsellers. Paperback writers were making a good living, and a number were upwardly mobile on the success ladder.

But dark clouds were forming over this way of life. A large segment of the reading audience had moved to the suburbs, where supersized bookstores were popping up like mushrooms in malls all over the country. However, bookstore shelves were not happy homes for mass market paperbacks, which had always been most effectively displayed face out on easily accessible revolving racks in grocery markets or drugstores. As chain stores drew fans away from those shops, mass market paperbacks were relegated to bookshelves in remote corners of the vast emporiums. And they were often stocked in hard-to-read spine-out mode, discouraging to shoppers who (despite the adage) were able to tell a book by its cover.

The configuration of those shelves lent themselves better to the trade paperback format. The term "trade paperback" originated in the 1980s with works of literary fiction published by houses like Penguin and Vintage.[17] They were a bit bigger than the standard mass market paperback: about 5×8 inches versus the mass market paperbacks' 4.25×6.87 inches. But in the 1990s, "trade paperback" came to connote softcover reprints of approximately the same dimension—"trim size" in publishing lingo—as the original hardcover, 6×9 inches.

A big advantage of trade paperbacks was that unlike their mass market cousins, they were not a monthly phenomenon and could therefore be kept in a store for as long as there was demand for them. Thus, trade paperbacks eventually became *the* format for all but reprints of bestsellers, driving another nail into the coffins of writers of mass market originals.

For another thing, computerized sales information now enabled publishers, wholesale agencies and retailers to better track the performance of categories and identify winners and losers among specific books and authors. Ominously, upon assessing these patterns, paperback distributors began asking themselves why they needed to employ commissioned jobbers when they could more efficiently and economically service bookstores and other outlets by shipping copies directly to the retailers. Yes, telemarketing would mean that the human element—the driver who knew which towns loved historical romances and which preferred crime stories, which adored westerns and which were big on science fiction—would be removed from the equation.

In time, whole fleets of drivers were discharged and the wholesale distribution workforce reduced to a fraction of what it had been in its heyday. One day, customers and store owners in west Texas or Nebraska or South Carolina woke up to discover that many books by their favorite authors were no longer being stocked in their local stores. When they complained, they were told to take it up with the distributor in Vancouver or some other faraway location reachable only by an 800 phone number.

A result of this shift in distribution patterns was the streamlining of the way retailers ordered books from publishers. Why pick and choose among thousands of titles that might sell only a handful of copies? Wasn't it better to follow the formula that worked so well at airports, ordering only the top fifteen or twenty bestselling books by branded authors like Nora Roberts, Robert Ludlum, John Grisham and Stephen King? The damage to certain genres was severe. It is entirely possible, for instance, that the loss of distribution to truck stops severely damaged the western genre, so beloved of truck drivers.

As paperback publishers awoke to the new buying patterns, they were forced to choose between star authors and those whose sales performance fell below a minimum level. At first, the triaging was restricted to marginal authors and genres, but as the last decade of the twentieth century progressed, the definition of "marginal" broadened to embrace any writer whose sales fell below an ever-ascending threshold of commerciality.

Paperbacks were also shedding their pulpy reputations. A number of western, romance and thriller authors built huge paperback audiences and achieved respectability, garnering serious reviews and evaluations in literary publications like the Sunday *New York Times Book Review* section. These writers were now being acquired (aggrieved paperback publishers said "stolen") by hardcover houses. The paperback houses not only lost their authors but had to pay a premium for the right to reprint the very writers they had published for years and years at affordable advances.

Hardcover publishers were far from thrilled about having to give 50 percent of those big paperback reprint advances to authors, as was stipulated in their contracts. They therefore cast acquisitive eyes on paperback companies. Out of mutual need was born the idea of merging hardcover and mass market paperback houses, enabling a single entity to acquire and exploit both formats and make more money on both. For example, in 1980, *Princess Daisy*, a steamy adult fairy tale of wealth, glamor and royalty by Judith Krantz, was auctioned by Crown Publishers among eight paperback houses, ending up at Bantam for $3,208,875 (over $13 million in today's currency), a record at the time.[18] Had Crown owned Bantam, it could have reaped the profits from both editions and probably paid less to the author. (Today, both companies are part of the Penguin Random House empire.)

The process of marrying paperback houses to hardcover ones was tumultuous, and among its many harmful by-products was the further debilitation of smaller trade publishers who had no mass market sister houses to reprint their books. Be that as it may, by the end of century, the amalgamations were locked in place. The paperback tail had successfully wagged the hardcover dog.

The rise of hard-soft publishers (as they came to be known) dealt a hammer blow to legions of paperback writers, for these consolidations had raised the cost, and the stakes, of acquisition. Whereas a paperback publisher formerly made a profit—and the author a living—on books that sold five or ten thousand copies, now the combine had to sell tens of thousands of hardcover copies and hundreds of thousands of mass market paperbacks. Not many authors fit that description. Those that didn't dropped out; those who survived the shakeout were relegated to the opprobriously termed "midlist," connoting writers competent enough to be published but not commercial enough to achieve breakout bestsellerdom.

The pseudonymous "Jane Austen Doe," a self-styled member of the midlist rank, lamented her status in a 2004 Salon.com essay:

> In the 10 years since I signed my first book contract, the publishing industry has changed in ways that are devastating—emotionally, financially, professionally, spiritually, and creatively—to midlist authors like me. You've read about it in your morning paper: Once-genteel "houses" gobbled up by slavering conglomerates; independent bookstores cannibalized by chain and online retailers; book sales sinking as the number of TV channels soars. What once was about literature is now about return on investment. What once was hand-sold one by one by well-read, book-loving booksellers now moves by the pallet-load at Wal-Mart [the then-current spelling] and Borders—or doesn't move at all.[19]

What the new breed of publisher—and perhaps the new breed of editor that worked there—failed to understand was that paperback originals had been a kind of vocational school for authorship. A writer could write dozens of books for a barely living wage until their skills matured and they were at last recognized and rewarded. Now, with that source of apprenticeship gone, where were seasoned professional writers going to come from? The new vocational school was the School of Hard Knocks, and the curriculum was, "Tough shit, Jack, you're on your own."

As the publishing industry entered the twenty-first century, book industry executives began requiring editors to produce elaborate profit-and-loss projections and other corporate-style analyses of the potential viability of books and authors. What was the sales performance of previous books? Did they "sell through" satisfactorily, or did returns cross the threshold of unprofitability according to the latest formulas devised by bookstore-chain number crunchers? The mantra of "the bottom line" was invoked *ad nauseam* at every editorial committee meeting, and editors were constantly reminded that "we can only afford to publish hits. If you can't project a big profit on a book, turn it down."

Whereas the founders of publishing companies were decisive and courageous individualists willing to roll the dice for something they felt passionate about, succeeding generations of management tended to grow more and more risk-averse. A kind of middle management mindset afflicted the decision-making process: It was safer to say no to a project than to risk one's job greenlighting a book that might flop and lose money. Fearful of incurring liability, editors stopped negotiating deals over the telephone. Though disputes had arisen from time to time out of phone negotiation, the prohibition—even to iron out wrinkles in a complex agreement—was frustrating to a generation of schmoozers. Everything had to be done by email.

Computerized financial projections were aided by new market-research applications. Now, within moments, editors could access author track records and sales statistics on previously published books. And all too frequently, acquisition committees relied more on performance metrics than on traditional but less quantifiable values like compelling storytelling or stirring prose.

And what about the authors? Were they attractive? "Mediagenic"? Did they have a "platform"—an organizational base, such as a hit television series or a chain of fitness centers, capable of promoting the sale of books? Whenever I pitched a book over the phone, I could hear the click-clack of fingers on a keyboard as the editor explored the author's website to see what the author looked like and how compelling was their site. "Yes, it's a good book," editors seemed to be thinking, "but is that all it is?"

More and more, the importance of traditional literary criteria took a backseat to "the numbers" and "the platform." Promising but modestly successful novelists discovered they could not get their second or third books published, and aspiring newcomers could not sell their books at all because they were not branded.

Faced with these grim options, authors resorted to increasingly frenzied measures to get published. Established novelists wrote under pen names to disguise the poor performance of their earlier books, or strove to produce blockbuster "breakout" novels long on sex, violence and plot, but short on craft and characterization. Without supportive publishers to carry them while they developed their talents over four or five books, new novelists focused on gimmicky concepts with log lines that could be pitched like movie scripts. Non-fiction authors plumped up their credentials or hired specialists to burnish their images and enhance their social media exposure. Others subsidized the purchase of large quantities of their own books to drive up their numbers. Literary agents were besieged by writers frantically seeking the advantage of representation by high-profile deal makers.

The age of paperback originals was drawing to a close, and as much as authors longed to cling to it, that was no likelier than a return to dial phones. A few venues for mass paper originals continued to thrive, mainly romance houses, which generated at least 25 percent, and by some accounts as much as 50 percent, of all paperbacks sold. But even romance specialists like Harlequin and Kensington, and Tor in science fiction, saw the market for originals beginning to evaporate. All but the biggest and most enduring books failed to make the cut. Steven Zacharius, President and CEO of Kensington Publishing Corporation, told me in 2023 that his company issued only eight mass paperbacks a month, down from thirty a few years earlier. In 2025, as we shall see, death notices for the format were posted in the industry trade publications.

But at the beginning of the twenty-first century, the pocketbook trim size had become yet another instrument of blockbuster publishing, especially for reprints of hardcover bestsellers. The printings were usually enormous, commonly over 100,000. And economies of scale—the bigger the printing, the lower the unit cost—enabled publishers to make a profit even with high returns, sacrificing untold numbers of trees to the Moloch of a prodigiously wasteful distribution system.

Although these upheavals ostensibly impacted print books only, they set the stage for the transformation of the twenty-first-century paradigm from analog to digital. An industry drowning in returns was about to be introduced to a new means of delivering books to readers via a system whose return rate was zero. Authors who had been frozen—or thrown—out of the print marketplace would soon migrate to the alternate universe of digital self-publication. Beloved works of literature that had stagnated in limbo were about to be revived and their authors rediscovered and celebrated and even enriched. Midlist and niche writers, working in obscurity and barely subsisting, would find their voices, attract fresh audiences, and enjoy unimagined streams of compensation.

Soon, but not quite yet. There were more tribulations ahead.

7. "PUH . . ."

(1999-2000)

Though patriotic saws we spout,
We farm e-book production out.
"Made in USA" a myth,
 Pick up the phone, you don't get Smith.
Continents and oceans spanning,
We outsource stripping, prep and scanning.
In Mumbai, Agra, Cooch and Mysore
Every proofreader is eye-sore.
Indians are making whoopee
Working for the Yankee rupee.

L OOKING BACK, I HAVE TO CONFESS that my decision to launch an e-book publishing company was ill-considered if not utterly reckless. Try as I have to reconstruct my frame of mind, I can't imagine what made me confident that the enterprise would succeed or even lift six inches off the ground. Perhaps it was a 2000 Andersen Consulting study reported in *The Miami Herald* predicting that e-books would enjoy $2.3 billion in annual sales within five years. (They didn't, falling almost $2 billion short of Andersen's projections, according to the Association of American Publishers.)[1] Or perhaps it was a line from Bill Gates's essay, "Content Is King": "Content is where I expect much of the real money will be made on the Internet, just as it was in broadcasting."[2] Like most delphic prognostications, however, that one was long on inspiration and short on specifics. For all my dreaming I had not thought everything through. Hell, I hadn't thought *anything* through!

In my defense, there were so many variables in the evolving industry that a coherent business model was not yet possible. Nevertheless, numerous elements really should have been nailed down by the time we got underway.

For instance, should our royalty be based on the list price or net? Most trade book publishers base their royalty on the list price of their books, that is, the one printed on the cover. A 10 percent royalty on a $25.00 book is easy enough to calculate. But I had observed that many retailers were heavily discounting books, making for a highly volatile pricing climate. It made far more sense to base our royalty on the net—that is, on what we actually received, which was standard for academic books.

And what should that royalty be? As former head of the agents' organization, I had strongly advocated 50 percent. That was easy enough to advocate, but now I was a publisher. How the hell could I make money giving half of our receipts to those greedy authors and their conniving agents? But I had committed to it and now I had to make it work. I turned to Gates again: "This technology will liberate publishers to charge small amounts of money, in the hope of attracting wide audiences." I knew that over time our production expenses would be amortized by a perpetual stream of revenue and we'd make a profit. The trick was to stay in business until profitability kicked in.

I accomplished this with the simple expedient of charging authors for the cost of conversion—creating digital texts from printed books—and production, at that time, $250 per book. The reasoning was that if authors are getting 50 percent royalty they should contribute to the expenses. However, rather than require authors to pay this charge up front, we offered to lay out those expenses and recover them from royalties. This made it painless, and most of them accepted the scheme without demur.

Those outlays were a form of advance, but the question arose whether we should pay actual advances against royalties like grown-up publishers. As long as we were reinventing publishing, we decided against that practice, hoping that the inducement of a bigger slice of the royalty pie would compensate for the absence of advances. It did for most authors, but unfortunately, for most agents, it did not. To an agent's way of thinking, a book deal is not legitimate unless it is bound by front money (a percentage of which is their commission). And many of them represented authors we were keen to publish. Alas, we had to defer that satisfaction until our company was "mature" (definition: had enough money).

Thus, our standard contract looked like this: We acquired worldwide English language print and e-book rights for a term of five to ten years. As we were not (at the time) capable of exploiting translation rights, we reserved those to the authors, as well as audio, movie and television rights. We paid a 50 percent net royalty on all revenue we collected and charged $250 for our production expenses, recoupable from royalties. One other innovation was quarterly issuance of royalty statements and payments, as opposed to semiannual statements that were standard trade book industry operating procedure. We figured that a no-returns business model would be more efficient than the convoluted system used by traditional publishers, enabling us to produce royalty statements faster (we figured correctly). This business model served as a template for other e-book publishers that eventually came into the business.

There were two attachments to our contract. The first was a guide to assist authors in the formatting and proofreading of their digital files; the second was a release form in which my literary agency waived its commission on my publishing company's revenues. The authors confirmed that they had had the opportunity to consult with an attorney or agent and were satisfied that their rights were protected by the contract.[*]

Many other issues caught me unprepared. Other than a layman's *Popular Mechanics*-type understanding of science, I had no technical expertise whatever. Nor did I possess marketing skills or e-commerce experience. I had no clue how to go about converting printed books to computer-readable digital files, sell them online, collect revenues or disburse royalties. I was a fast learner but should have been better prepared before I undertook this enterprise. As a result of my inexperience (or let's face it, ignorance), some lessons were administered to me in short sharp shocks. Which explains why my eyebrows are in a permanently raised position.

Be that as it may, on January 5, 1999, the New York State Department of Taxation and Finance issued Certificate of Authority #13-4049399 for E-Rights/E-Reads, Ltd., an S-Corporation.

E-Reads was to be the publishing branch, E-Rights the agency for helping authors and agents clear digital rights to their books. (Credit for both names goes to my wife, Leslie.) We managed to trademark E-Reads, but our trademark

---

[*]A specimen contract, with some explanatory commentary, will be found in Appendix A.

application for E-Rights was turned down by the government because the term was too generic. They were certainly right about that, as the phrase soon entered common industry jargon.

A handful of partners teamed up with me and contributed to the company's capitalization, but with one invaluable exception, they dropped out one by one and, over time, sold their shares back to me. The exception was Eric Golden, a financially astute friend of our family who put money behind his faith in our future. It was not just as an investor but as a consumer that Eric recognized a need that was not being fulfilled. "I was drawn to the company because e-books offered a convenient way to read wherever you are, given the ability to access potentially unlimited books anywhere in the world and while traveling; plus the backlight feature allowing reading in tough places such as on planes, at night."

The biggest asset E-Reads had, and the one I was going all-in on, was a starter set of more than a thousand books. These were genre novels whose e-rights my agency had either recovered over the previous decade or managed to withhold from publishers, titles that I knew legions of fans yearned to see back in print. This collection included former bestsellers, award-winners and masterpieces in their fields. Among them were fantasy and science fiction by authors Harlan Ellison, Fritz Leiber, Dan Simmons, Greg Bear, R. A. MacAvoy, Greg Keyes, George Alec Effinger, Dave Duncan and John Norman; romance stars Janet Dailey, Jennifer Blake and Laura Kinsale; crime novelists Jim Thompson, Barbara Parker and Richard S. Prather; and horror author Ray Garton.

Though viable e-book readers were years in the future, I knew that early adopters were fanatical enough to read these works on desktop or laptop computers, handheld PDAs, or any other device that reasonably served to display text on a screen. I was also confident that developers of e-book reading devices would covet this bonanza.

In short, books were my killer app.

There was just one problem: They didn't belong to me.

Legally they belonged to the authors. As their agent, I had terminated their publishing contracts, which canceled Richard Curtis Associates' status as agency of record. Legally speaking, they were no longer my clients. For the purpose of granting me the right to publish their books, they were independent authors— in sports terms, "free agents."

I had recovered and warehoused the rights, counseling my clients to wait patiently for the fledgling industry to mature. They expected that when the time came, I would do what agents traditionally do with unexploited rights: license them to a publisher and take the usual agent's commission. But when I looked around, I was unable to identify any publisher that I felt confident entrusting our e-books to. That was when I decided to publish them myself.

But wait a minute—didn't that make me a buyer?

It certainly did, which meant I was sitting on a conflict of interest the size of the Matterhorn.

As former President of the Association of Authors' Representatives (AAR) and one of the authors of its Canon of Ethics, no one understood conflict of interest better than I. With respect to that transgression, the canon stated: "Members shall not represent both buyer and seller in the same transaction." The key phrase was "the same transaction." If my agency took a commission on royalties paid to authors by my publishing company, it would clearly be a "same transaction" violation for which I would—rightfully—be subject to censure or even expulsion.

I did not remotely harbor any such design, but because I had launched E-Reads so abruptly, I had not had time to discuss it with my authors. This was a serious and potentially fatal misjudgment. However unsophisticated an author may be, they understand the difference between an agent and a publisher, which may be summarized as "Agent Good, Publisher Bad." If they believed that their agent had gone over to the dark side—or worse, was double-dealing—I could lose not only my publishing company but my agency as well.

I got on the phone and didn't get off for a month. If I could make the case that both my agency and my publisher functions could coexist, I had a shot at retaining my authors' loyalty. I explained that I had scrupulously taken every measure to separate publisher from agency in form, function and financing so as to avoid even the merest hint of conflict. Although my agency would continue handling their non-e-book business such as movie or translation deals, no commission would be taken on monies payable to authors by E-Reads. As a further assurance of good faith, I encouraged them to engage an attorney or another agent to negotiate their contracts with E-Reads.

In my favor, they knew (if I may say so) that I was a straight shooter, a fierce authors' advocate and a person who followed the highest ethical standards in the conduct of his business. For years, they had heard me extolling the glories of the coming E-Book Revolution, and now I was going to deliver on my vision. I demonstrated how they stood to make at least twice as much money per sale as they made in traditional publishing. Best of all, their long-out-of-print books would be brought back to life.

To my great satisfaction, almost every author accepted the dual arrangement and reaffirmed their trust. A handful chose to wait and see how the new industry shook out. And there were a few skeptics who believed that the only good book was a print book. These holdouts would eventually come around— but that was okay, I could wait. I had my starter set of books. E-Reads was in business.

The AAR—well, that conversation didn't go quite as benignly. Though there was no overt opposition, I could sense the members' unease. The distinctions I drew between the functions of agent and publisher were clear and logical, and most of my colleagues grasped it on an intellectual level. But on the visceral—or should I say olfactory—level, the plan did not pass the test. An agent who was also a publisher was outside their experience or value system, and it just didn't smell right. A number of my colleagues pointed out a variety of ways that the wall between agent and publisher could be breached by a person of dubious integrity. As for my own integrity, some of them said that no matter how honorably I ran my business, there were enticements that would tempt a saint.

Whatever the merits, a quarrel with the organization would have been public and damaging, and I did not want to inflict it on these men and women of goodwill, many of whom were my close friends and veterans of many a wrangle with publishers. I visited the chairman of AAR's ethics committee and tendered my resignation. He accepted it with a mixture of regret and relief.

In the fourteen-year history of E-Reads, I'm happy to boast that the question of conflict of interest never once came up. But interestingly, within a few years after I started the company, a number of other literary agencies started e-book publishing companies. Aside from one agent, I don't know how the agents reconciled the two conflicting functions. But Joshua Bilmes, founder and President of JABberwocky Literary Agency, did it simply and elegantly. Unwilling to sell out-of-print titles to traditional publishers, who took an unacceptably high 75 percent share of e-book revenue, Bilmes decided in 2010 to produce and publish them out of his office and take no more than an agent's commission after recouping production costs. Unlike my operation, Bilmes did not separate the roles of agent and publisher. JAB Books was simply a commissionable service performed by his agency. He also used e-books to solve a common problem facing agents: what to do with the third book in a trilogy if the publisher didn't want to publish it. Bilmes simply added it to his own e-book list. "My business is being a literary agent," Bilmes told me.

With our cache of books at last in safe harbor, I was able to focus on the countless other challenges and perils that had to be navigated. To address them I made two important decisions. The first was to assemble a staff with expertise in the various competencies necessary to run the business—programming, production, design, distribution and accounting among the most salient. This was far easier said than done, as I could not yet afford a team of experienced specialists in those disciplines, and I sometimes had to ask people from my agency to pitch in. But I could not risk using amateurs to develop a website or guide us through the technical challenges of production and what is now called asset management. Thus, in 2002, we hired Michael Gaudet, a savvy technician with publishing experience, to superintend those tasks, and for seven years he managed our operations with a steady and able hand. We could not have gotten through without him.

Another critical rule was to farm out all tasks that were too complex or costly to perform in-house. We therefore subbed out every job we were not absolutely confident we could handle successfully on our own. "When in doubt, farm it out" became our motto. Though it meant sacrificing a goodly portion of our revenue, it was far cheaper and safer than buying and managing complex programs or equipment. Better to pay people who know what they're doing than waste money trying to do it yourself. Or as the saying goes, "Fifty percent of something is better than one hundred percent of nothing."

Take the matter of book conversion, the almost alchemical transformation of printed page to machine-encoded text readable on a screen. Technically the process is known as Optical Character Recognition. OCR in that embryonic era required a number of complex and delicate steps. First, the book's cover had to be stripped. Second, the sewn or glued spine had to be chopped off. Third, the "running head" (title, author, page number, usually at the top of every page) had to be sliced off, too. Then the loose pages had to be fed into an optical scanner, which took digital photos of them. Print on paper in, digital text on screen out. Then the scanned text had to be proofread.

Except for the proofreading, every step of the conversion process required special equipment and technical skills far above our pay grade, and the process was a prime candidate for the assignment of burdensome tasks to outside operators. Needless to say, OCR is far faster and easier today.

Where to find a conversion house? It just so happened that for a brief period, 1997–99, Barnes & Noble was working with one in order to expedite production for e-book publishers. The publishing behemoth had created the barnesandnoble.com website under Steve Riggio, younger brother of the company's potentate Leonard.* As already mentioned, Barnes & Noble had invested in the Rocket eBook, arguably the first true e-book (about which more to come), so they were obviously on the hunt for content to populate it with. Ken Brooks, B&N's Vice President, Digital Content Division, took a shine to our company and connected us to an offshore conversion outfit. Brooks was knowledgeable, wise, generous with his counsel and well connected, and on more occasions than I can count, he lit our way out of many a dark corner, including a number we had painted ourselves into.

The conversion house not only scanned our books but produced the covers as well. Cover art for e-books at that time was a work in progress, to say the least. Designs were minimally pictorial and came in such hues as muddy red, muddy blue and muddy green, which presented as murky rectangles when reduced to the thumbnail dimensions recommended for book websites such as Amazon's. But I did not complain. The alternative was to purchase cover art from publishers at prohibitive prices, and on our budget, anything over $1.50 was prohibitive. So we elected to use the conversion house's standard-issue cover art. Later a skilled and innovative designer named Andy Ross joined our team and from that point on we designed our covers in-house, utilizing inexpensive stock photos that we could alter to taste, then add title, author, and other key elements. Andy's designs were attractive and eye-catching, and he was particularly skilled in fashioning covers for series books that required variations in color or design from one volume to the next.

We also intended to produce printed books but were informed that we needed to place bar codes on their covers (e-book covers do not require them.) Who knew? I'd read a million books and scarcely noticed that every cover had a scannable bar code. I learned that it contained important information like title, publisher, format, price, weight and the unique ten-digit identifier (changed in 2007 to thirteen digits) called the ISBN (International Standard Book Number). So, after purchasing a block of ISBN numbers from a company called Bowker, we produced bar codes and placed them on the cover of every print edition.

---

*After 2003 the company left the e-book business until it introduced the Nook e-readers six years later. Leonard Riggio died in 2024.

And placed them not just anywhere. If we wanted them to be properly picked up by scanners, we had to leave a blank area of precisely 3.625 × 1.25 inches in the lower right-hand corner of the back of the book and drop the bar code there.

An adjunct of producing covers was writing jacket (or cover) copy, the description of a book's story and characters. The easiest thing would have been to use the text employed by the original publisher. The problem was, I didn't think we had the right to do so. Although their copy may not have been copyrighted, I reasoned that it must belong to the publisher that paid someone to write it. We therefore wrote our own jacket copy—or paraphrased someone else's.

Our conversion house offered proofreading in its suite of services. But I had serious reservations. Proofreading is a highly labor-intensive task, requiring a sharp-eyed and meticulous individual to scrutinize every word, usually with a second copy of the book on hand to clarify garbled text. Early OCR products were rife with typos caused by anything from dirty scanner lenses to yellowed paper to poor typography in the original volume. For instance, with certain fonts, "off" might be misread by the OCR machine as "oll"; "wear" could be mistaken for "wean"; "clip" could turn into "dip" and "horn" into "hom." Given the pathological aversion to typos that I shared with all right-minded editorial folk, I demanded 100 percent clean texts. That may sound fanatical until you do the math: a 75,000-word book that is 99.9 percent error-free will contain 75 typos. Given that I get heartsick over a single one, 75 would send me to an early grave. In time, we hired our own proofreader.

The production process was equally stultifying. The RTF file—Rich Text Format, a word-processing format that was the basic framework of our e-books—must be reviewed page by page by a designer to make sure it reads seamlessly. "Once a book gets scanned," our Production Manager, Nate Fernald, explained, "it tends to lose all of its formatting with the exception of single line breaks. [A line break is added white space between two paragraphs, sometimes graced by asterisks.] And line breaks must be clearly delineated to prevent scene shifts within a chapter from running into each other. When we get a file back from scanning, I have to flip through the physical book, page by page, comparing it with the file to see if there was any formatting lost, such as centered text, indented text, extra line breaks, etc."

The staggering monotony of this process will explain why I granted Nate one day off every week. He was beginning to exhibit classic symptoms of PTSD.

One more station of torment: we had to provide metadata—vital book-related information—to the retailers. It includes cover image, ISBN number, language, territorial rights, suggested retail price, publication date, brief description and other details and data, such as a BISAC code, a nine-character cipher summarizing a book's genre, topic, and theme (BISAC is an acronym for Book Industry Standards and Communications). Retailers demanded pages and pages of metadata definitions, specs and tolerances, all in fine print. And each retailer had different requirements or the same requirements but listed in a different order.

Okay, we now had our e-books: formatted with proofread texts and covers. In short, we had everything we needed except for a means of selling them. We had a website to display our wares; now all we had to do was sell them from that site.[*] It did not take very long to conclude that that was out of the question.

To vend our books online would have required an e-commerce website, webmaster, credit card processing service, liability and fraud protection insurance, customer service, tax licenses in every state and every nation abroad, and many more features that were far beyond our skills and light years beyond our budget. And that was just for e-books. As it was our plan to sell print books, too, we would have to print, bind, warehouse, pack, label and ship them. There was no way on the planet that we were capable of or interested in doing this.

It looked like we were screwed before we began. I was reminded of a serialized novel I had read as a boy. The hero has fallen into a twenty-foot-deep pit with sheer sides. He has no tools and there is nothing to grab hold of to pull himself out. How on earth is he going to escape certain doom? I counted the days until next month's issue. At last it arrived. I tore open the wrapper and breathlessly turned to the next installment. It began: "With one mighty leap . . ."

My escape from these predicaments was not quite as heroic, but fortune did smile on us in the form of two recently minted operations.

The first was an e-commerce website called Fictionwise. It had been launched in 2000 as a partnership between two brothers, Scott and Stephen Pendergrast. When I came across their website, I noted that they were selling digitized short stories (for 49 cents!) by some well-known science fiction authors, and they had figured out how to sell them online. "Our advantage," Scott

---

[*] The current website ereads.com is not related to the publishing company that is the subject of this book.

explained, "was always that we were a technology company. Stephen and I had started companies around digital content (in our cases, online training) before working together on Fictionwise. Our direct retail competitors were often from publishing. So we could be nimble and add more and more features with our five-person company that our competitors couldn't match. We had personal reading recommendations long before Amazon, for example. E-Reads found a way to outsource this work brilliantly, but others spent millions hiring full pro-gramming teams. Hard to make that money back in the early years. It was big news when these companies folded so quickly. It made analysts question the entire market. It made all of us wonder where this was all going—so it wasn't necessarily positive in our minds when competitors dropped out."

Obviously, the Pendergrasts had what we needed and vice versa. We in-vited these smart and earnest young men to our offices and made a deal to sell our e-books. Though we eventually added many more vendors, our relationship with Fictionwise was especially warm, stabilizing and mutually profitable.

Recently, Scott Pendergrast reminded me just how mutual the relationship was: "I am not sure you understand how important E-Reads was to us. As you know, we sold mostly science fiction short stories when we started. When you offered us the chance to sell full-length novels by many of the great science fic-tion authors of all time, it opened the door to what the e-book industry could become. It made us think bigger, and it gave us faith that *you* had faith in the model. It was no small thing."

Above all, Fictionwise served as proof of concept: *If you have the content, they will come.*[*]

Okay, the problem of distributing e-books was solved. But what about printed books? These were a vital component of our business plan. However keen early adapters were to read e-books on a desktop or laptop screen, many fans were dedicated to conventional print and hungry to hold in their hands hard (and inexpensive) copies of long-sought works. This strategy proved cor-rect: For years after we launched E-Reads, 50 percent of our revenue came from printed books. But without a means of producing or distributing them, we were up the creek without a paddle—until Lady Luck waved her wand a second time.

---

[*] In 2009 Barnes & Noble acquired the Pendergrasts' company for $15.7 million, and these trailblazers deserved every penny of it.

On October 2, 1990, Xerox unveiled a printer called the DocuTech 135 Production Publisher, billing it as "ushering in the print-on-demand era." In a retrospective blog published twenty-five years later, Xerox's Gregory Pings explained that the machine—a beast some eleven feet long and weighing more than a ton—combined three technologies: high-resolution scanning, laser imaging and xerography. "Within a decade of its introduction," declared Frank J. Romano, Chair of the Rochester Institute of Technology's School of Print Media, "most of the black-and-white printing work in the United States had shifted to DocuTech, profoundly changing the nature of the printing industry and moving it into the digital age." Randall Hube, manager of Litigation & Strategic Technical Services in Xerox's Intellectual Property Operations, put it more succinctly, calling it "a print shop in a box."[3]

Xerox displayed this masterpiece at a BookExpo America, the annual convocation of publishers, booksellers and, increasingly, tech people. It happens that John Ingram, Chairman of Ingram Content Group, attended that event. There he beheld the DocuTech 135 printing a single copy of a book and saw the future of print-on-demand publishing. It took some time to develop a business model in which the customer selects a book from the publisher's or retailer's website and prepays for it by credit card. (Electronic retailers came to be dubbed "e-tailers.") Then and only then is the book printed, packaged at the plant and drop-shipped to the customer's address. "Essentially," Pings wrote, "the work process of printing documents was flipped on its head and fundamentally changed forever." Up to then the sequence was "Print and distribute, then hope to sell." Now, thanks to POD (as the technology was acronymed) the new model was "Presell, then print and ship."

The concept of printing one copy at a time and printing it only *after* being paid for it was almost impossible for most publishing people to get their heads around. But John Ingram got it, foresaw a market and created LightningSource, a company dedicated to print-on-demand publishing. In 2000, only two years after the company printed its first POD, it printed its millionth.

I, too, got it and seized on it as the solution to E-Reads' quandary. Indeed, I seized on it as the solution to the book industry's quandary and evangelized a world to come where consignment distribution was a thing of the past. I imagined miniature POD presses installed in bookshops and libraries. Future generations would look back at twentieth-century publishing and wonder, "How did people live that way?"

My dream came true, at least for a while, notably with a POD machine promoted by Jason Epstein, former editorial director of Random House, who, in 2003, cofounded the company that made the Espresso Book Machine.[4] A number of bookshops installed it, leading Epstein to predict it would make traditional printing techniques obsolete. As much fun as it was for store customers to watch the Espresso produce a book in front of their eyes, however, it did not take off as Epstein had hoped. (In 2022, following Epstein's death, his cofounder Dane Neller told *Publishers Weekly* that Espressos are still active in several locations, and he believes there is still a future for what he called "an ATM for books.")[5]

In any case, we began feeding text and cover files to Lightning. Ingram had made a deal with amazon.com to sell PODs on its website. Every month Lightning provided us with a statement of sales for each book accompanied by a check. The process of producing, selling and reporting was reduced to a minimum of steps, and payments were issued on the dot every month. This was as close to "add water and serve" as you could get, and Lightning became a treasured relationship throughout our years in business.

How did the finances work? We started by asking how much royalty we wanted to receive for each title. Suppose we said $3.00. To that figure we added Lightning's printing charge, calculated on the number of pages. All our books were paperbacks 8.5 × 5.5 inches, a trim size between trade and mass market paperback. On the average, a book of 300 pages cost (in those days) about $6.00 to print. Adding that figure to our $3.00 target royalty gave us a wholesale price of $9.00. Amazon (in those days) took a 100 percent markup on sales. We therefore doubled the wholesale price to arrive at the book's retail price of 18.00. In other words, Amazon bought the book from Lightning for $9.00 and sold it for $18.00. Out of the $9.00 Amazon paid to Lightning, Lightning recovered its $6.00 printing cost and sent us our $3.00 royalty, half of which we kept and the other half we remitted to the author.

Even though printed books are analog artifacts, I've always regarded PODs as tangible e-books, texts projected onto paper versus e-book texts projected onto screens. As satisfying (and profitable) as PODs were, however, my vision from the beginning had been e-books, and in 2000 Amazon made a move that dramatically furthered that goal: They launched a dedicated e-bookstore website. They called it e-Books—not a very imaginative name, but I didn't care. The formats were PDFs and Microsoft Reader. Ingram made a deal for Amazon to distribute its e-books—including *our* e-books!—making LightningSource a major contributor to our bottom line.

I was not familiar with the term "outsource" when I started E-Reads, but the moment I heard the word I embraced it, and it became the guiding principle in running my newly hatched business. Attributed to management consultant Peter Drucker, who coined it in 1989, the Grammarist website describes it as a portmanteau word meaning "to contract jobs or tasks that were previously provided inside a company . . . in order to reduce overhead and increase efficiency, productivity, revenue and profitability to remain competitive."[6] Although I initially regarded outsourcing simply as a prudent business strategy, the more I thought about it, the more I realized that the practice had larger—indeed, profound—social and philosophical significance.

The train of thought started when I asked myself: Of the numerous tasks involved in producing e-books and operating my company, were there any that absolutely required centralized management of our operation? I ticked one after another off my list and concluded that the only indispensable one was the obligation of our staff to work on the premises. Teleconferencing was in its infancy, and on our parsimonious budget, purchasing that equipment was out of the question.[7] We were shackled to the server in our office, which contained our shared database. We liked one another and worked well together, but I could easily imagine that one day an effective form of remote audiovisual communication would be devised that functioned as simply and easily as the telephone. And when it did, was there any reason for employees to work in a central office?

In a flash of insight—call it Epiphany 1.2—I grasped the essential difference between the universe in which I had functioned for my entire career and the one I was plunging into: In the analog world, the locus of one's business is a street address; in the digital world it's an IP address. It followed that a business could not only outsource goods and services but labor as well. A business establishment would no longer have to function in a fixed location but rather in a network of nodes linked to a central server, making brick-and-mortar headquarters irrelevant.

Today, post-pandemic, remote networking is accepted wisdom. But in the year 2000 a vast dispersion of the labor force from office to home was almost unimaginable. Yet I imagined it, and more.

These speculations were magnified when another term entered tech parlance around this time: "disintermediation." This two-dollar word originated in the banking industry, where it applied to consumers cutting out—"de-middling" as it were—the banks that traditionally handled their financial transactions, and instead invested directly in stocks, bonds or other securities.

The process was not limited to banks. The growing power of home computers and the diminishing centrality of offices, stores and other physical establishments created the potential for disintermediation in every field of endeavor. Indeed, with each new outsourcing the centrality of such an establishment dissipated until the only palpable thing left was the server. Agencies servicing travel, brokerage and real estate found themselves squeezed as buyers and sellers discovered the value—or savings—in direct relationships with each other.

It was not just agencies in the literal sense that stood to be disintermediated. Any retail business could be regarded as an agency between manufacturer and buyer. Shops and department stores were therefore vulnerable to consumers who could directly purchase goods and services at the source. Their transactions, executed by a server, were a preview of online shopping to come and its cataclysmic impact on physical stores.

Traditional publishing is a quintessentially analog industry (one of our staff described traditional publishers as servers in skyscrapers). Despite some digital applications developed at the end of the twentieth century, it was still heavily dependent on laborers working in physical buildings producing hard goods delivered by fossil-fueled vehicles to a network of physical retail shops. The financial structure of the industry reflected this fact. Bookstores took 50 percent of a book's retail price, and publishers took most of the other 50 percent. Given that publishers and bookstores underwrote every process, did all the heavy lifting and bore the brunt of huge overhead, they were able to justify their entitlement to the lion's share of the revenue. But that overhead made them prime targets for disintermediation, as Jeff Bezos recognized when he launched amazon.com in 1994. In actuality, amazon.com is a server disguised as a retail store, a server so powerful and efficient that the company is able to effect huge savings by furnishing books to consumers at prices far below its competitors. The magic of Amazon is the opacity of the process. All that underlying infrastructure is invisible to you when the box with the smiley swoosh arrives at your door.

E-Reads was designed along similar principles: a writer, a reader and a server. Virtual texts produced by a small labor force working out of cheap quarters and distributed by computer. Our publishing company was the server under my desk. Though we couldn't completely extricate ourselves from overhead, what savings we generated were passed along to our authors. For example, traditional mass market paperbacks in 2000 were typically priced at $5.99. The author earned between 6 and 10 percent—$.36 to $.60—royalty per copy. An e-book at the same list price, however, would earn a royalty of up to $1.50: list price less Amazon's 50 percent discount less our company's 50 percent share. (In 2010, Amazon changed its basic royalty commission to the current 30 percent.)

As things progressed in the new millennium, outsourcing and disintermediation would have disruptive if not calamitous consequences for the publishing industry—and for literary agents. But in 2000, the thrill of blazing new trails eclipsed any reservations we pioneers might have harbored.

So—we were launched. People started calling me a player. Ha! Here's what a player I was. The following May I attended Book Expo in Manhattan's Javits Center. In ordering my badge, I decided to identify myself as the President of E-Reads instead of the Curtis Agency. As I did the crawl through the exposition's aisles I bumped into an old friend who squinted at my badge. "E-Reads—what's that?"

"I'm a puh . . ." I stammered. "A puh . . . a puh . . ."

For my entire career publishers had been the Enemy, and now it was impossible for me to utter the detested word *publisher.*

So yes, I was a player, a one-hundred-percent genuine puh.

8. THE BIG SLEEP

(2000)

The book trade suffered a recession
In every genre but confession.
Because we had to have our sleaze
The mills destroyed a lot of trees.
No publisher could lose its shirt
On books that dished up tons of dirt.
Celebrities whose sheen had tarnished
Revealed in print The Whole Unvarnished.
Literary agents hondled
Deals for stars who had been fondled.

W ITH ALL THE BUILDUP, all the giddy excitement of pioneering, all the fanfares and revolutionary pronunciamentos, with all the dazzling technology and fancy acronyms, you would imagine the new millennium would usher in the Glorious E-Book Epoch with parades and fireworks. You certainly wouldn't have dreamed that the e-book industry came close to collapsing.

In March 2000, Nobel Prize-winning economist Robert J. Shiller published a pessimistic study of overvalued tech stocks. He took his title, *Irrational Exuberance*, from a speech Federal Reserve Chairman Alan Greenspan had given in 1996. "How do we know . . . when irrational exuberance has unduly escalated asset values," Greenspan had asked, "which then become subject to unexpected and prolonged contractions?" Publication of Shiller's book came at the zenith of the boom, as exuberant investors raced towards the precipice from which so many dreams were about to plummet.

For years, venture capital and starry-eyed optimism had inflated stock values to levels unsupported by underlying capital or prospective revenues. *Publishers Weekly*'s Jim Milliot and Calvin Reid reported that one company, Reciprocal, "had raised some $80 million, while netLibrary went through more than $110 million. Questia has raised more than $130 million."[1] On April 10, 2000, the Nasdaq Composite Index began a weeklong fall of 25 percent in the Internet sector. A number of business reversals and scandals aggravated the woes of investment bulls.

Seven months later, the decline of World Wide Web stocks sank to 75 percent off their highs.[2] Among the losers was Amazon, whose stock fell below $10 from a high of more than $100. The 9/11 attack on the World Trade Center in 2001 and the ensuing Iraq war struck yet another blow to an economy already reeling from the high-tech recession. The stock market floor looked like an abattoir: By 2002 Internet-based tech stocks had lost $5 *trillion* of capitalization.[3]

E-book companies and supporting services like data conversion fell like ten-pins. "DigitalGoods, Versaware, Reciprocal, WizeUp, et al.," Milliot and Reid recounted. "And several start-ups that aimed to supply digital content to consumers also closed their doors: Audiohighway.com, Booktech, Contentville.com, MightyWords.com." By 2001, some two thousand e-publishing workers had been laid off; some four thousand jobs in traditional publishing and retailing were also cut.

Two big publishers that had ventured into e-books folded their operations. Random House ditched its e-book venture, AtRandom.com, in 2001. Another victim was iPublish.com, a premature version of the self-publishing phenomenon that a decade later would flourish. Created by Time Warner (now Hachette) in 2000, iPublish "was conceived as an accessible outlet for emerging writers, using the Internet not just for distribution but for democratizing the publishing process," the Associated Press noted. "It encouraged opinions and feedback from readers as well as other writers on submitted manuscripts". A year later it was gone. "The market for e-books has simply not developed as we hoped," lamented Laurence J. Kirshbaum, head of Time Warner's trade division.[4]

A hush fell over the industry as if the spell out of *Sleeping Beauty* had put everyone to sleep. It was followed by a hiatus so long that those who had formerly expressed alarm that e-books spelled the doom of paper now wistfully recalled them as a fad, like Beanie Babies: "Hey, remember e-books?"

Depressed spirits were temporarily revived, in March 2000, by Simon & Schuster's publication of Stephen King's novella *Riding the Bullet*. It was the first part of a novel in progress called *The Plant* and billed as the world's first mass market e-book original. Utilizing e-merchandising designed by an outfit called SoftLock.com, the story sold so heavily on its day of release—400,000 units—that it crashed the servers of SoftLock, amazon.com and Barnes & Noble. King self-published subsequent installments, but they sold far less successfully.[5]

With so much in the e-book world still unsettled, *Riding the Bullet* was more of a singular event than the initial temblor of the long-awaited paradigm shift. Its principal value was to raise consciousness about the commercial potential of the new medium. But even under the byline of one of the world's most popular authors, *Riding the Bullet* could not kick-start a cold engine. In fact, its performance was not as spectacular as it first seemed: sales of the second volume were less than half of the first.[6] The world lapsed back into lassitude and denial and sought comfort in the safe and familiar harbor of printed books.

Aside from the death or crippling of e-book developers, the fact was that a clearly defined e-book business did not yet exist in the year 2000. All the elements were out there but not fixed in place. The Great Sorting Out—the fusion of technologies, the consolidation of procedures, the homogenization of countless standards—was still in the future. Publishers, authors, agents and, above all, consumers groped for a distinct model in an atmosphere of confusion, uncertainty, skepticism and even hostility toward the viability of this new medium. In a businessinsider.com post Michael Mace, CEO of Cera Technology and former executive at Palm, Apple and Silicon Graphics, wrote that "the market for e-books and e-book reader devices utterly failed to take off the way most observers expected in 2000." Among the reasons he gave were that there was not yet a truly viable reader; nor were there enough digitized books, the existing ones were too expensive, and poor marketing had failed to motivate consumers to abandon printed ones.[7]

A significant aftermath of the 1998 NIST conference was the realization that everyone was on a different page. Each e-reader had its own unique format, making consumers wary of committing to one brand for fear they would not be able to migrate to a new model or transfer their books if a better device came along. Customer service had difficulty keeping up with all the fixes and upgrades introduced by the various manufacturers. This situation was intensified by their policies known generically as DRM. That stood for "Digital Rights Management," a chaste euphemism for *roadblock*. DRM formats and codes were proprietary, that is, not interchangeable with other e-readers nor easily adaptable to upgrades. If the industry was to succeed, order had to be imposed on this chaos: a one-size-fits-all format shared by everyone involved in the medium.

The first step in achieving this goal was to identify the areas of disorder and create a framework of rules and standards to which all participants in the arena were required to adhere. However competitive companies may have been with each other, they would have to agree on a measure of uniformity in a variety of hardware and software applications, specifications and procedures. The key word was "interoperability." Simply stated: If a format works on your device, it's gotta work on mine, too.

With guidance and encouragement from NIST, a consortium was created called the Open eBook Forum, dedicated to creating an "industry standard for authoring reusable content for e-book devices." This standard did not prevent developers from innovating and improving their products; they had plenty of leeway to outperform each other in any one of a number of features like shape, weight, speed, brightness and capacity. But at the very least they had to do it on the same playing field as their rivals, the dimensions of which would be prescribed by the OeBF. "Comprehensive specifications that achieve the goal of simultaneously supporting both interoperability and innovative functionality are absolutely critical if electronic publishing is going to deliver on its promises," said Allen Renear, chair of the OeBF Publication Structure Working Group.[8]

There were countless specs to be wrestled to the ground. To take one example, in November 1999, a team of five e-book developers, including members from SoftBook and NuvoMedia/Rocket eBook, issued a thirteen-page draft document replete with code ("Open eBook™ File Format 1.0") addressing the urgent need for content providers to squeeze large, complex electronic files into a single file to transfer electronically to users.[9]

In 1992, an Internet standard known as MIME (Multipurpose Internet Mail Extensions) had been established enabling content providers to attach to emails a variety of content, such as images, audio, video and text. But now, seven years later, e-book files had grown unmanageably big. The goal therefore was to create "a means by which content compression may be accomplished" and the content transmitted in "a single transportable file." The vehicle for accomplishing this compression was called a "gzip."[*]

Compression and portability were just two of the mass of standards that had to be sorted out. But sort them out the working group did and brought forth the Open eBook Publication Structure. The key word was "Open"—the format was not protected by DRM. "The OEB Specification provides a single common format for authors, editors, publishers, and content owners who want titles to be readable by a variety of electronic publishing systems and reading devices," said Bob Bruce, executive director of the Open eBook Forum. (At E-Reads, we had always eschewed DRM.)

I attended a number of OeBF gatherings, and though a lot of the jargon was often over my head, I got the gist and was able to translate it into lay lingo. More important, I connected with a number of developers with whom we eventually did business.

The OEB standard prevailed for the first five years of the new decade, but owing to a flood of research and development advances and the influx of more players, the original consortium "reinvented itself in 2005 with a broader remit," in the words of that organization's Executive Director, Bill McCoy. OeBF became IDPF: the International Digital Publishing Forum.

Recognizing that the early OEB file extension was no longer adequate, the IDPF set about creating a new one and in 2007 announced EPUB as the new standard. EPUB is defined by the World Wide Web Consortium as "a means of representing, packaging and encoding structured and semantically enhanced Web content . . . for distribution in a single-file container." It was and is supported by most e-book readers.[†]

But all that was still to come. It was 2000 and our humble little start-up had to make do with the modest resources at hand. We hadn't been hurt by the bursting of the dot-com bubble, because we had no venture capital to lose except my own. I can joke about it now, but it was a matter of no little concern to my accountant. I had invested a quarter of a million dollars in the first two years,

---

[*] A gzip is capable of compressing only a single file, whereas the more familiar "zip" compresses multiple files and archives them into one file.

[†] While unique ISBN identifying numbers are all but universal in the book industry, Amazon created its own unique identifier called ASIN: Amazon Standard Identification Number.

with little to show for it. Even my wife, who had encouraged me to follow my e-book dream, looked nervously at the descending balance in our bank account. When I assured her that I'd been tracking industry statistics and e-book sales had doubled in the last year, she quipped, "Yes, from one dollar to two dollars." Actually, industry sales were in the single-digit *millions* of dollars, and yes, they were growing year by year. The graph would rise steadily to $1 billion in 2020. I had faith that our own growth would parallel the industry's.

We did grow, though on a laughably lower level. As a matter of fact, our first revenues, minuscule though they were, triggered a crisis that nearly capsized us.

One morning in 2000, the mail brought an envelope containing a sheaf of royalty statements and a check, our very first! I can't remember the source, but I do remember my heart swelling as I gazed upon this bounty. Against all manner of travail, E-Reads had made money! I understood the shopkeeper's tradition of framing the first dollar paid to them in their new establishment. And "first dollar" was almost literally true: Our retailer had sold but one or two copies of each of four books, and the check was not much larger than ten dollars.

A few days later we received envelopes from other retailers. More statements, more money. Yay! I handed the pile to our bookkeeper, Liced Cintron, instructed her and left her to the task of processing the data and issuing our own statements and payments to the authors.

About two hours later, following the sound of muttered imprecations, I looked in on her. She was still entering the information into her computer.

"What's the problem?" I asked.

"This is taking me all day," she complained. "These statements are all on paper."

She explained that she had to manually transfer all the information into a computer database. Because each company stated its royalty information differently, there was no standard format. "If it takes this long to enter a few sales, what am I going to do when we have hundreds?"

My heart sank. She was absolutely right. If we could not efficiently process royalties we would not survive another six months. There had to be an IT solution. I made some calls and was referred to a sharp programmer by the name of David Marlin. He came to our offices, and I explained the problem. He studied the statements, echoing our bookkeeper's grievance. "These reports are on *paper*!" After a few minutes, he declared, "I think I know how to fix it."

He asked me to provide a list of authors, titles, vendors and royalty rates from which he could create a central repository. Then he asked me to list the six or seven vital categories of royalty information we needed to report to authors.

Generally speaking, publisher royalty statements contain dozens of pieces of information, such as the accounting period covered by the report; the formats, publication dates and list prices of each edition; the author's royalty percentage per edition; the current period's sales; the lifetime cumulative sales; and more. It is not necessary, however, for all of this detail to be reported to authors. David's idea was to screen out the less relevant categories and funnel the ones we needed into the essential subheadings that I provided to him. "Think of twenty railroad tracks approaching a terminal that has only six or seven platforms," he explained. "Using Excel, we will import those twenty 'tracks' of vendor data and funnel them into a relevant few." He smiled. "I'll need a couple of months to develop this."

"A couple more months," I groaned, "and my bookkeeper will jump out the window."

"No. You're going tell your vendors, *'No more paper!'* " I channeled Faye Dunaway as Joan Crawford in *Mommy Dearest* shrieking at her daughter, "No wire hangers *ever!*"

"They must furnish all statements as Excel files or you won't accept them," David stressed. "That will expedite the bookkeeping until the cavalry arrives."

And it did. Over the next few months, he developed an elegant system that he called "Royalty Tracker," a masterful royalty processing and reporting tool for the digital age. When we began issuing royalty statements, I included a covering letter to authors relating our achievements and the state of the e-book industry, and it became a quarterly custom.[*]

I thought that a lot of publishers and agents would be interested in acquiring his application, so I proposed to use my connections to target them. One thing led to another, and David and I decided to partner on the venture.

Garnering input from his other clients, he refined and expanded Royalty Tracker. But as gratifying as the operation's growth was, between E-Reads and Royalty Tracker and my agency, I was spreading my attention too thin. David wanted to go his own way, anyway, so he offered to buy out my share, and I

---

[*] A sample letter is reproduced in Appendix B.

agreed. His company, Metacomet Systems—now cloud-based—has evolved into one of the leading royalty management companies in the publishing industry. But it was inspired by our bookkeeper's threat of defenestration.

9. SHIFTING SANDS

(EARLY 2000s)

> *E-books wakened from their funk.*
> *Emailed books? No longer junk!*
> *When editors and agents shmooss,*
> *As like as not they'll introduce*
> *A jargon-laden nomenclature*
> *Like none Linnaeus found in nature.*
> *"DADs" and "DOIs" and "PODs" and "Digits,"*
> *"RAM" and "ROM" and "Gigs" and "Widgets."*

I WOULD NOT WISH A FINANCIAL CRISIS on anyone, but from a purely selfish viewpoint, the recession following the dot-com collapse was an unexpected blessing.

For one thing, the atmosphere at E-Reads was calmer. With a couple of years of experience under our belts and the support of a network of like-minded pioneers, we shed the anxieties that usually beset start-ups. Though there were many surprises, disappointments and even shocks ahead, none would rattle us the way our initial confusion and uncertainty did. Our team was more experienced, and whatever challenge arose, at least one of us remained cool and calm and knew how to handle it.

We settled into a more organized and businesslike operation. The assembly line and production process became more efficient, and we cultivated relationships with many new vendors coming on board. Our strategy of declining to operate a retail website was paying off; by midsummer of 2001 we had distribution deals with Barnes & Noble, Fictionwise, Gemstar, Palm, Adobe, and some eight other retail sites. These companies possessed the resources to do the heavy lifting of order fulfillment, customer service and security, allowing us to focus on what we knew and did best: aggregating content, converting it to digital files and distributing it through third parties.

As a result, we started making a profit in 2002, netting $35,000 on revenue of $253,000. Modest though it was, it represented a confidence-building proof of concept.

I realized that some of the technical knowledge I had absorbed in my digital sojourn could be effectively applied to improve the productivity and profitability of my literary agency. I've mentioned how our submission process was revolutionized by the discovery of a means to attach digital files to emails. Although this development was ostensibly devised to solve a very simple problem, it contributed more than any other to propelling our business into the twenty-first century.

Take the configuration of office space. We had a whole room dedicated to storing sturdy boxes specifically designed to contain book manuscripts, plus stacks of folders for proposals and padded envelopes for postal submissions. There was a long table for packing and mailing with a postal scale and stamp machine. Our office walls were lined with vast bookshelves containing as many as two dozen copies of each book we represented. We required so many in order to service submissions to movie studios and foreign publishers or to bestow on editors to showcase our clients.

The introduction of email submissions wiped out the need for all this space and consigned those artifacts to history. Even printed books became little more than window-dressing. Movie scouts and story editors, agents, publishers, translators, talk show hosts—all were satisfied with or even insisted on Word or PDF files. Publishers and agents now used their newfound office space to make room for other tasks or more personnel. Or they could move to smaller quarters unencumbered by so much—well, *stuff*. As gratifying as it was to behold shelves colorfully adorned with book jackets, the savings of time, labor and money created by digital submissions balanced those satisfactions and often outweighed them. Even for a small agency like ours, the cost of shipping submission and production copies of hardcovers overseas totaled thousands of dollars annually. Now we were able to move that item off the red side of our ledgers.

And don't forget the savings in time. Multiple submissions of books we were auctioning used to take hours typing up submission letters, packing up manuscripts and delivering them to the post office (invariably on days when twenty customers were lined up ahead of us). The same task now took an hour or two at most.

Beyond these practical considerations was something unsettling, even (if I may say so) cosmic. For it wasn't just *our* business that was transformed, but business in general. The severance from concrete objects offered liberation from *place* and set society's feet on a path that would lead two decades later to the migration of labor from business office to home, where one could function and even thrive with little more than a laptop and cell phone. But it was not just a geographical migration; it was psychological: Society's mindset had begun to shift from material to virtual.

This shift was an immense boon to authors, publishers and agents, and cyberspace soon coruscated with vibrant and dynamic websites. Yet as time went by, I began to grow uneasy about unintended consequences.

Under pressure to present their wares more and more attractively, authors began to develop web design skills, write penetrating jacket copy, develop marketing strategies, engage professional photographers and videographers for glamorous author photos and entertaining promotional videos, and (after working so hard to purge their writing of adjectives) puff up their bios to heroic dimensions and herald them with eye- (and sometimes ear-) catching displays. Websites grew glitzier and more elaborate, replete with biographical and bibliographical pages, pictures and videos, blurbs, review quotes, awards and honors, sales rankings, and assorted decorations and frou-frou, reminiscent of hot rods festooned with trinkets and ornaments.

I had speculated on whether book editors were too sophisticated to be taken in by all this pageantry. In a blog entitled "Watching Books," I wrote:

> It is not unreasonable to speculate that a lifetime of exposure (if not addiction) to media—indeed, to *multi*media—may have compromised editors' ability to judge books on their own merits. Rather it is tempting for editors to judge them in the context of entertaining audiovisual displays. As successive generations accustomed to being diverted by watching, rather than by reading, enter the editorial workforce, impatience with printed text is demonstrably increasing, as we can see in the sharp decline of print newspapers and magazines.

> Books require a commitment of time and attention that we either don't have or aren't willing to give. The temptation to skip or skimp is strong. One editor confessed to me, "I tend to scan manuscripts on screen rather than read them the way I do a printed text." We must therefore ask ourselves whether, instead of *reading* books on screen, we are *watching* them.

The sciences of metadata and market research provided sales information and audience preferences, furnishing publishers with tools for quantifying a book's significance or an author's popularity. One heard terms like "branding" and "platform" with growing frequency as the preferred criteria for judging submissions. All too often the first question at acquisition meetings was not "How's the book?" but rather "How many followers does the author have?"

It was growing clear that editors' attention was shifting from the quality of books to the sex appeal of authors, the splendor of their platforms and the size of their followings. Books now had to be more than books and authors more than authors to quicken editorial pulses. A first novelist unsupported by strong social media exposure and armed with nothing more than a manuscript, no matter how good, was at a serious disadvantage.

It was so much safer to sign up a Somebody, however dreadful, than a Nobody, however wonderful. How much easier it might be to sell a collection of a superstar's spaghetti recipes than a brilliant, insightful book by someone you never heard of. I remember seeing a copy of Bill Clinton's *My Life* on a friend's bookshelf. Proudly she showed me his autograph. "The President called me darlin' when he signed it," she gushed. "Did you read it?" I asked. "No, but he called me darlin'," she repeated. For her, Clinton's book was a trophy whose contents were irrelevant. Were books becoming celebrity souvenirs?

Opportunities for unknown authors to get into the limelight were on their way, however.

Advances in digital technology provided an opportunity to streamline publishing industry news, and in 2000, an Internet-savvy book publisher named Michael Cader applied his skills to transform the discourse of a generation of book people. As Christopher West Davis of *The New York Times* describes, it started when Cader began emailing "a free copy of a breezy, irreverent newsletter called *Publishers Lunch* to 21,000 agents, editors, publishers, writers, movie scouts, book store owners and librarians." His goal, he announced, was to become "the electronic equivalent of the old-fashioned publisher's lunch, a place where all of us can swap completely unsubstantiated but dead-on stories, news, deal information, and insights."[1]

What started casually (Cader called it a hobby) became more serious, better organized and profitable, as subscribers paid for news and information that was both cogent and entertaining. In time, he teamed up with a web developer to produce Publishers Marketplace, a storehouse containing such content as deals, industry news, book reviews, bestseller lists and analyses, a job board, classified ads, rights offerings and the *Publishers Lunch* archive. "Arguably," writes publishing consultant Power, "one of the three most important research tools of my time in publishing with Bookscan and Amazon."[2]

Cader expanded *Lunch* into what he termed a "deluxe" edition, analyzing industry developments, deals, mergers and acquisitions, personalities, promotional campaigns and even lawsuits. Although some of the same material could be found in *Publishers Weekly*—the venerable (founded in 1872) flagship of the trade—book people either preferred Cader's colorful presentations or consumed both publications for a comprehensive understanding of what was happening in the business. In 2024, he announced dramatic AI-driven innovations "powering all kinds of cool searches of our data."[3]

But it was his innovative formula for reporting book deals that revolutionized industry communications. Until "Deal Lunch," few deals were reported unless they involved prominent authors, record-breaking advances, a headline-making story or all three. Announcements of small, conventional deals for books by unknown or barely known writers and involving modest amounts of money were consigned to press releases by author, agent or publisher, but few escaped the gravitational pull of obscurity.

"Deal Lunch" offered the opportunity for authors to find a place in the sun and for every deal to be writ almost as large as reports of eight-figure blockbusters. However, the privilege came with a condition, and one that sounded like a game-show challenge: *The transaction must be described in a single sentence.*

To a "civilian" (as we sometimes refer to nonpublishing people), this condition may seem easy enough to satisfy, but for a race of articulate if not verbose literati, encapsulating their precious books was agonizing. It meant stuffing the five W's of journalism—Who, What, When, Where and Why—into one line and doing it alluringly and eloquently. The task of compressing a rhapsody of hundreds of words into a one-liner is as challenging as Saturday's *New York Times* crossword puzzle, and publishers and agents whiled away many an hour whittling words, sentences and even paragraphs down to a single sublimely compact utterance. To aid struggling submitters in their homework assignment, Publishers Marketplace created a questionnaire with such fields as Author, Agent, Acquiring Editor, Imprint, Territory, Price Range, "Author Bio (maximum 150 characters)" and "Book Description (maximum 500 characters)." "Deal Lunch" applies an algorithm (perhaps AI) to this information and produces a formidable single sentence that might read:

> Flyfot Prize-winning author of *1001 Holiday Recipes for Triscuits*, Constantine Puptent's *I Married an Emperor Penguin*, illustrated by Jonquil Nightsoil-Kickpleat, in which a fraternity pledge stows away on a ship he mistakenly believes is bound for Sumatra but after a hilarious series of contretemps fetches up in Atka Bay in Antarctica with an amorous female Emperor Penguin before he is rescued by a Japanese whaler manned by a troupe of circus acrobats (more hilarious contretemps ensue), pitched in the vein of *To Kill A Mockingbird* meets *Wuthering Heights* meets *Atlas Shrugged* meets *Go the Fuck to Sleep,* to Nigel Framster at Lackluster House, in a pre-empt for publication summer 2085, by Simon Bungwart at Smuddle & Earpod Agency (World rights).

This amalgam of log line and elevator pitch appealed to a young, impatient, tech-savvy, hit-the-ground-running cadre of publishing people who were completely at ease sending and receiving information electronically, a generation that took to tweeting and texting as soon as they were old enough to pick up a cell phone. Cader's formulation became standard procedure for pitching books, inspiring this bit of doggerel I composed in my annual poem for *Publishers Weekly*:

> We learned to format Deals for Lunch,
> Our hype compressed to one long munch,
> A style devised by Michael Cader
> With log-lines longer than a seder.

10. SLOUCHING TOWARDS KINDLE

(2000-2005)

Thus in frenzied syncopation
Proceeds the trade's consolidation.
Scores of famous names of yore
Have since succumbed to corporate war
Or publish books with but a semblance
Of their former independence.

T O MANY FOUNDING EVANGELISTS of the e-book movement, the first four
or five years of the new century were disillusioning. Gung-ho enthusiasm
had dampened in a climate of uncertainty and skepticism. After pep-rallies like
the NIST conference, activity became eerily quiet. To all appearances, e-books
seemed more of an ephemeral novelty than a necessity like VCRs or
smartphones (or printed books, for that matter). Growth was painfully incre-
mental. The International Digital Publishing Forum reported wholesale e-book
sales of $5.8 million in 2002, $7.3 million in 2003, $9.6 million in 2004 and $10.9
million in 2005.[1]

A host of companies and organizations worked with little fanfare to advance and refine the technology, digitize books and forge the standards on
which a solid industry would one day be erected. Among these diehards was my
pint-sized operation, as my faith in the value of content showed signs of paying
off. After running at a stomach-turning deficit for a couple of years, we made a
small profit. About 50 percent of our revenue came from print-on-demand paperbacks, demonstrating either the enduring strength of print or consumer

dissatisfaction with existing e-book readers. However, owing to heavy investment, the next few years returned E-Reads to stomach-turning deficits before the arrow on our chart turned permanently north.

The sources of this munificence were numerous, for we were liberal if not promiscuous in our partnerships. If you were an e-book distributor and solvent, we were happy to take your money (all contracts were nonexclusive). A review of vendors who contributed to our coffers in the fourth quarter of 2002 includes Fictionwise, Peanut Press/Palm, LightningSource (both e-book and print on demand) and Rocket eBook/SoftBook, the latter two now under the corporate banner of Gemstar. Alas, that company would soon succumb to economic hardship and disappear from our roster. But it was fun while it lasted, and more important, their gadgets projected an ineradicable image of how e-books should look, feel and function. Rocket and SoftBook were in the right church but, unfortunately, the wrong pew.

Three newcomers appeared on our roster of distributors that year: Content Reserve, Baker & Taylor and netLibrary. They were all in the business of purveying e-books to libraries.

I have to confess that despite decades in the publishing business, I had given little thought to traditional libraries and zero thought to e-book ones. The concept of lending physical books was simple, but when I applied the same rules to digital books, I got a headache.

I had a million questions. How do you lend an e-file to somebody? How do you get it back when they're finished reading it? How do you get it back if they *haven't* finished reading it? Can a library make multiple copies of a file and lend them simultaneously to everyone who requests it? How do publishers make money selling e-books to libraries—or rather, how do publishers keep from *losing* money selling e-books to libraries? Now that we were doing business in this field, it behooved me to learn how it worked.

Traditional libraries purchase one or more copies of a book from publishers. Under the rule called first-sale doctrine, a library has the right to lend a print book to an infinite number of borrowers. But because a print book is a tangible object, it can be loaned to only one borrower at a time. If it's a hot bestseller, the library may buy multiple copies. When interest cools off, however, the library will be stuck with a surplus of copies taking up shelf space needed for new books. For a book of normal general interest, a single copy usually suffices, and borrowers have no choice but to wait in line until it is returned by previous borrowers. If there is an unexpected surge in demand—say the book is selected by Oprah's Book Club—the library may have to order more copies. For librarians working with perpetually inadequate budgets, which are usually subsidized by state or municipal funds, balancing these tensions is a never-ending juggling act. But the process is fairly logical.

The introduction of e-books defied logic, however. E-files cost a fraction of a printed book. As it cost no more for a publisher to upload fifty copies of an e-book than to upload one, or to allow a library to make fifty copies of that one file, the traditional lending library concept of "one copy, one user" was rendered meaningless.

From the viewpoint of librarians and borrowers, the technology was a blessing. The savings would free libraries to satisfy numerous patrons simultaneously and immediately without incurring the prohibitive cost of buying (or overbuying) multiple printed copies. For publishers, however, it was the stuff of nightmares. The prospect of an infinite number of users accessing infinite copies of one e-file was nothing short of traumatizing. There had to be ways to limit use.

And there were.

In a *New Yorker* article, "The Surprisingly Big Business of Library E-books," Daniel A. Gross pointed out that "the first-sale doctrine does not apply to digital content."[2] This means that e-book publishers can set restrictions on licensees limiting the number of uses or the length of usage time. Another copyright law— 17 U.S.C. 106(3)—recognizes the right of the copyright owner, such as the publisher, to make and distribute *multiple* copies of a work. In other words, it is illegal for a library to make a dozen copies of a single e-book file that it licenses from a publisher. But if that publisher *sells* a dozen copies of that file to the library, the library may legitimately distribute all dozen—but each to only one borrower at a time.

Unfortunately, as crazy-making as these perplexities were, the passage of time did not resolve them. In 2011, for instance, HarperCollins imposed a 26-loan limit on all its e-book titles.[3] In 2012, Random House increased its e-book pricing to two to three times the list price of the same books in hardcover.[4] Seven years after that, in 2019, the issue remained red-hot. Macmillan imposed an embargo on availability of e-books in libraries until two months after publication of the print edition. In response, *Publishers Weekly* reported, at least one library system, King County, Washington, "has decided it will no longer purchase embargoed e-book titles from the publisher."[5]

In a 2012 article in *American Libraries Magazine*, two library executives listed numerous choices facing librarians, all of which were flawed in one way or another, as you can probably surmise:

- *The metering model:* Pay-by-the-download.
- *The rent-to-own model*: After a fixed time or number of loans, the library owns the e-book.
- *The bookshelf model*: The library pays an annual fee for a set of e-books selected by publishers or distributors. At the end of a year, it vaporizes and another set is delivered.
- *The embargo model*: E-book availability is blocked until print edition sales have slackened.
- *User fee model:* Users pay an annual fee for a library card, in effect subsidizing the purchase of books.[6]

Although Content Reserve, Baker & Taylor and netLibrary approached their missions differently, the business models were fundamentally the same. Of these, Content Reserve was the oldest, biggest and most resourceful.

Launched in 2000, Content Reserve was a dedicated digital platform for public libraries. Its parent company, OverDrive, had been founded in 1986 by Steve Potash, a true pioneer who grasped and applied digital technology in its infancy, converting print documents and then books, and storing them on CD-ROMs. Potash had envisioned a market for library e-books and given a lot of thought to the complex and delicate relationships among publishers, libraries and book borrowers. He concluded there was a way to satisfy all three and make money doing it. He did it so successfully that he added music, magazines and videos to the content that was purveyed to libraries.

Potash was universally admired and respected, and served as President of the International Digital Publishing Forum (IDPF) for many years. He eventually sold his company for so much money, over $400 million, I wanted to cry.[7] As my late father once said to me, "I think you're in the wrong racket."

Although the dot-com collapse had thrown a wet towel over the e-book industry for four or five years and the sales numbers in the early aughts were pretty modest, e-developers and publishers forged ahead, sustained by a steady stream of positive news. Even absent a satisfactory handheld reader, e-books were selling and being read on laptops, desktops, smart phones and PDAs. Whenever I felt discouraged—or my accountant cleared his throat—I pointed to the IDPF's quarterly sales chart showing year over year growth for almost every quarter (see chart below). In my November 2003 letter to authors accompanying royalty checks for the third quarter of that year I noted that "Retailers reported 48% percent growth in e-book units sold and 32 percent growth in e-book revenues. Publishers increased unit sales by 65 percent and revenues by over 30 percent."

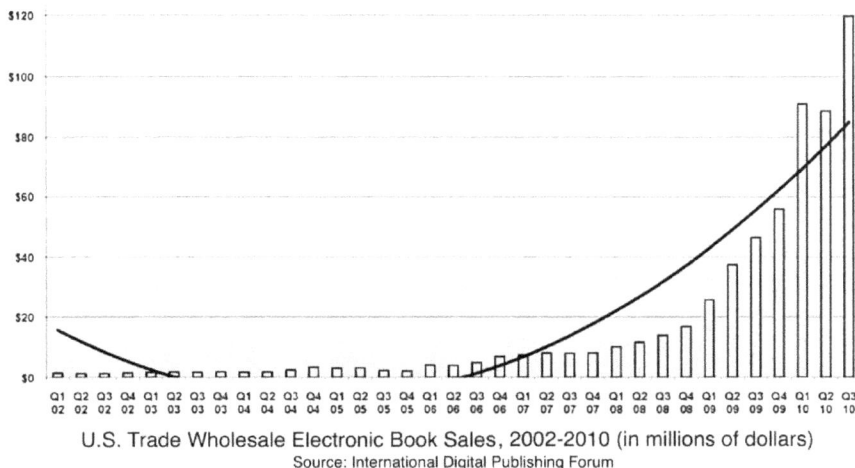

U.S. Trade Wholesale Electronic Book Sales, 2002-2010 (in millions of dollars)
Source: International Digital Publishing Forum

Our own growth roughly paralleled that of this graph, but because we reinvested revenues in production, a number of years in the first decade showed losses and would continue to do so until 2009. But that was all right. We were riding an updraft. In a memo to authors in mid 2004, I reported that "E-Reads now has more than 2,000 books for sale or in production, exceeding the e-book lists of most mainstream publishers, including Simon & Schuster, HarperCollins, and Random House. A growing number of literary agencies have placed books by their authors with us. In addition, we have contracts with several publishers to convert their books from conventional print to digital edition." One of these, Kensington, put 100 titles into our hands.

Weary of gloomy media coverage of the industry, I thought it a good idea to remind book critics that despite struggles and setbacks, the e-book trend was definitely upward. In a shamelessly boosterish diatribe in *Publishers Weekly* entitled "Bullish on E-Books," I wrote:

Those of us making money in e-books, delivering thousands of downloads every month, paying royalties to authors and publishers, have to wonder what planet these pundits are on. How could perceptions be so completely at odds with evidence that the e-book industry is healthy and pretty much where it was projected to be on growth charts?

In their haste to gloat, observers have been less than balanced in their analyses of the failures . . . while thriving, well-managed companies are ignored or their achievements minimized. Analysts accustomed to measuring best sellers on a scale of millions of copies belittle or ignore the pride (and profit) that e-book publishers take in a few hundred downloads of an e-book.

Perhaps our industry is out of touch with the consumers who are actually purchasing, downloading and reading e-books: the early technology adopters, the fans thrilled to find cherished books back in print, the young people for whom reading on handheld devices is commonplace. How long are we going to endure skeptics telling us nobody wants to read a book on a screen, when in truth thousands are paying to do so every day? In their haste to badmouth e-books, have critics forgotten the advantages of which they sang when e-books were first introduced—their convenience, versatility, economy? Those factors haven't changed. What happened to all those encomia about the benefits of e-books to students, the elderly, the visually impaired? And has anybody asked authors how they feel to see even a few dozen copies of their out-of-print books selling to a new generation of readers?

As long as I was holding the bully pulpit, I decided to throw in a scolding:

It's begun to look as if a lot of editors, agents, and even authors are half-hoping the new business will fall on its face. Why would they wish such a thing? Are they so afraid of being displaced by the new technology that they would prefer to live with the status quo? But what a status quo! A trade book publishing industry that pulps one copy out of every two it prints; that has operated for decades on profit margins under 5 percent; that authors complain has become hostile to literary endeavor; that readers complain has given them less content and less pleasure. Can we afford to wish this new enterprise ill when (talk about failure!) the past few decades are littered with the cadavers of conventional publishers merged, acquired or crushed by an antiquated and appallingly wasteful business model?[8]

Rereading it, this peroration shocks even me in its candor. But I don't take back a word.

Simon & Schuster's success with Stephen King's e-book *Riding the Bullet*, however brief, inspired that company to dip more toes in the digital water, and other mainstream houses began venturing too. By 2004, there was enough activity on the Open eBook Forum to generate an e-book bestseller list.[9] On it were such commercial authors as Dan Brown, David Baldacci, James Patterson, Patricia Cornwell and Stephen King. The Bible (New International Version) made it onto the list, as well as some reference books like Oxford University's *Electronic Pocket Oxford English Dictionary & Thesaurus Value Pack*. I was especially glad to see genre fiction represented there, in particular, science fiction, as evidenced by the appearance of Greg Bear, Kristine Kathryn Rusch, Elizabeth Moon, and a *Star Trek* novel by Keith DeCandido. Unit sales were not disclosed, so it's impossible to say how many copies constituted a bestseller. I would guess low thousands, but in my experience, even a few hundred units sold in a concentrated period of time could register on the list.

Pioneering e-book originals was Baen Books, and in time, Baen began distributing some of our backlist titles. Another popular genre, romance, was represented on the OeBF list by Stephanie Laurens and Lisa Kleypas. We had counted on genre fiction appealing to early adopters, and this bestseller list confirmed we'd put our money on the right horse. (Be it noted: *The New York Times* did not create its own e-book bestseller list until 2011.)

Although a number of e-industry brothers- and sisters-in-arms fell during this turbulent period, the survivors formed the foundation of a strong and enduring community, veterans of a prolonged and hard-fought battle. E-Reads was beginning to be recognized as a leader in the field, not so much for the volume of its business as, perhaps, the volume of its founder's opinions, for I was incapable of passing up any opportunity to air them. *Publishers Weekly*, in its Y2K issue, cited me as one of "Eleven for the Millennium," profiling publishing industry individuals who have "embraced the convergent nature of the new millennium."

"We have to break the existing models and cross some lines," I said, "or publishers and agents run a very real risk of becoming irrelevant to the process."[10]

The June 26, 2000, issue of the *New York Observer* reminded its readers of my Nostradamic prophecies: "I have been a prophet for 20 years," I said. "I predicted that the publishers would cannibalize each other and we would be left with a handful of traditional publishers. I predicted the invention of electronic delivery of information and text that would make traditional publishing processes irrelevant.

"You know what they call a prophet who says, 'I told you so'?" I added. "A pain in the ass."

At least I didn't take myself completely seriously.

The most entertaining coverage was a March 2001 article by Chris Allbritton and Sara Nelson in *Inside Magazine,* with the delicious title "Publishing's Grumpy Old Visionaries." Interviewed were myself, Jason Epstein, publisher and print-on-demand cheerleader, and John Brockman, agent for cutting-edge science books. Although the serious intent of the article was that you *can* teach old dogs new tricks, the humorous slant was the advanced years of the subjects, whom they teased as "e-geezers." "They were once wunderkinder," the sub-headline said, "but today—with a combined age of 200—they're the wunder-menschen of the Brave New Book World."

Though some would call it ageism today (and if anything makes me grumpy, it's ageism), I took it as good-natured ribbing and even joined in the fun. When Epstein said, "I think people will not want to read on screen as much as they want to read in book form," and Brockman grumbled, "Using my Palm and all these devices you're talking about, there's something about them I just don't like," I retorted "You guys are too old!"[11]

11. THE ROAD TO E-DAY

(2005)

"No one reads," said Apple's Jobs,
"Atrophies your frontal lobes.
Video is where it's at.
Stuff your e-books in your hat."
When market share began to dwindle,
Jobs paid grudging heed to Kindle,
Then cashed in on th'ebook bonanza
With an iPhone app called Stanza.
Ah, Steve, hypocrite lecteur!

I N APRIL 2005, AMAZON ISSUED an intriguing press release: They had acquired MobiPocket, a French company. Mobi had developed software for downloading and reading books on computers and handheld devices.

The bulletin did not make much of a stir in book trade papers but to Bezos-watchers—certainly to this one—it was an unmistakable sign that Amazon was developing an e-book reader. For some time, he and his engineers had been tracking the progress of research and development in the e-book field, and we now know that the company had begun development of the Kindle in 2004.[1]

Despite the paucity of digital book content, it was plain that the space was still wide open for the right handheld. Desktops and laptops were not handheld; cell phones were handheld but not very good at carrying e-books; PDAs were handheld but you had to squint to read the screen; tablets were handheld if you held them with two muscular hands. The closest thing had been the Rocket eBook, but aside from a variety of functional difficulties, Rocket's parent company Gemstar had gone out of business. Given Amazon's grip on the book industry and access to countless titles, a well-designed, easy to use, book-dedicated device would sweep competitors off the table. Unless someone beat Amazon to it.

Someone did.

It happens that in 2004, Japan's Sony Group Corporation, one of the world's biggest and richest manufacturers of consumer electronic products, came out with an e-reader called the Sony Librie. For a couple of years, distribution was limited to Japan, but in fall 2006, a version dubbed the PRS-500 was released in the United States under an exclusive contract with Borders bookstores. The original Librie's functions were navigated by a keyboard, but for the PRS-500, Sony replaced it with a little pair of wheels for menu, settings, dictionary and page-turning. In 2007, distribution of the Sony was expanded to other outlets besides Borders.

Compared to future bells-and-whistles-loaded models and to competitive e-readers still over the horizon, the 500 was a pretty modest little gadget. E-books had to be loaded from a computer via a USB port. The device had a six-inch-diagonal screen and an internal memory of a mere 64 megabytes. That's a paltry capacity compared to today's behemoths, but even if some of those megs were devoted to the device's operating system there was sufficient storage to carry a handful of books. (Calculations vary from one-half to two megabytes for a 300-page, non-illustrated book. Michael Kozlowski, in the article "Documentary—The entire history of Sony e-readers and e-notes," estimates 80 titles.)[2]

But of greater significance, the Sony boasted a game-changing display feature: E Ink.

As I explained earlier, E Ink was developed in the 1990s by MIT scientists. It is a matrix of capsules containing electrically charged pigments, negative for black, positive for white. When an electric signal is fired into the matrix—say, the word "book"—the capsules flip to black within 50 to 250 milliseconds to form the word "book" on the screen. E Ink text is completely legible in sunlight and, unlike liquid crystal displays, uses negligible battery power, meaning weeks or even months of life before recharging.

The Sony E-Reader gave a vigorous boost to E-Reads' finances, and the growth of our revenue from that source parallels the growth of the e-book industry during that same period:

2007:	$3,785.00
2008:	$7,223.00
2009:	$24,832.00
2010:	$42,036.00

Sony's release did not upset Jeff Bezos as much as one might imagine. It was one of many e-book gadgets his engineers examined but found wanting. What most intrigued them was the Librie's use of E Ink, a feature they incorporated into the Kindle. But the competition lacked two key components: an abundance of content and a means of delivering that content wirelessly. The slow speed, cumbersomeness and inefficiency of cable delivery were growing intolerable, but new means of transmitting information and files now made wireless transmission a reality—as long as you could stuff it into device no bigger or heavier than a book.

It was with those goals that engineer Gregg Zehr, formerly of Apple and more recently of Palm, was approached by Amazon in November 2004 to discuss a secret pet project cherished by Jeff Bezos and ramrodded by Senior Vice President Steve Kessel. Cautiously (for Amazon was almost paranoid about having its plan stolen), Kessel explained that three forms of content—books, movies and music—were "ripe for being converted to digital," as Zehr explained in an interview with John Hollar for Computer History Museum.[3] Zehr said to himself, "'Oh, please not a digital book, please not a digital book.' Because in my mind the digital book had been tried."

But yes, it was a digital book. Of those three forms of content, Amazon had by far the biggest advantage with books.

That meant Amazon had to get into a completely unfamiliar field: manufacturing products. And that deeply concerned Zehr. "That was one of the first questions I asked Jeff. I said, 'Why on earth would Amazon want to get into the hardware business?' It's razor thin margins. It's a tough business. To get your first product out the door, it might cost 50 million bucks." But the more Zehr thought about it (reinforced by Bezos's enthusiasm, optimism and wealth), the more he saw the possibilities. Inventing an e-book reader would be a start-up, but unlike previous start-ups, this one was backed by tons of money and access to an unlimited stockpile of books.

Zehr came on board, and the boss set him up with what engineers call a "skunkworks"—a semi-autonomous development team—in Silicon Valley, called Lab 126.

The challenges that Zehr faced were numerous and seemingly insuperable. He enumerated them for Hollar:

> You start to realize, like, oh, yeah, there's a lot of infrastructure we had to go and build out—those business relationships, those technical connectivity issues, the wireless, the hardware, battery life issues, the software issue of on-store commerce or on-device commerce, straight from the device without any intermediary. Plus, another part of the vision was if you ordered the device, when it comes to you, it already comes pre-configured knowing your Amazon credentials, so you can just turn it on. It wakes up, it provisions out to the network, it goes and presents your credentials to the Amazon store in a secure way. And you're able to, within just a few minutes of turning the device on, you're connected and able to download content.

The biggest technical hurdle was that the device had to be wireless. Zehr hoped Bezos would not press for this feature, but he soon discovered the boss's rule: "If you had three meetings with Jeff and he kept bringing up the same issue, it was like, better pay attention." So Zehr's team found a path to wireless that included engaging a carrier (Sprint) that was as hungry as Amazon to innovate.

Technical issues aside, Bezos told Zehr that "the actual hardest part of the Kindle development team was to get the content owners comfortable with going to digital." (Needless to say, E-Reads had no trouble on that score at all.)

But what about consumers? What if they weren't comfortable with going to digital? Zehr answers the question with an anecdote. When the Kindle reached beta mode, the Lab produced a number of sample units, bugs and all, and handed them out to a focus group. When the participants finished their critiques, the team asked them to give back the devices. Their response: "There is no way I'm giving this back." Amazon took this to be an augury of public acceptance. How accurate was the augury, we shall soon see.

When they at last had a prototype, Bezos joined them for a weekend "sequestered at an undisclosed location," writes Dan Karlinsky, Amazon's Senior Content Creator and self-styled "storyteller." They spent several intense days downloading and reading e-books and brainstorming, testing navigation, readability and all the other properties that they hoped would make Kindle the sine qua non of its peers and competitors. When all the bugs were squashed and wrinkles smoothed, they were ready to roll out the Kindle. "Originally," Kessel says, "I told Jeff (Bezos) it would take us about 18 months to build the Kindle and we could do it with a couple handfuls of folks. It took us three-and-a-half years and a lot more than a couple of handfuls of folks."[4]

Critically important to Amazon was that their gadget must be devoted to a single purpose "rather than a multipurpose piece of hardware that might create distractions," writes Karlinsky. As artless as that statement may sound, in light of the multipurpose devices in development at Apple, Karlinsky's comment can be taken as a planting of the flag. The ensuing debate between advocates of purely reading-dedicated handhelds (i.e., the Kindle) versus those favoring such "distractions" as movies, music, audiobooks, emails, websites, graphics, text messages, video and still photography (i.e., the iPhone and iPad) was to become a matter of the fiercest contention.

Given Apple founder Steve Jobs's "epic sense of possibility" (in the words of Laureen Jobs's memorial tribute to her husband), a dedicated e-reader was low priority and would occupy a small corner of the vast network of Apple applications that would soon sweep over the world of personal computing.[5] Besides, Jobs was skeptical about the future of e-reading or any other kind of reading. In a January 15, 2008, *New York Times* interview with John Markoff, Jobs opined that the recently released Kindle "would go nowhere largely because Americans have stopped reading. It doesn't matter how good or bad the product is," he pronounced, "the fact is that people don't read anymore."

Jobs could scarcely have been unaware that even with a list price of $399 (about $600 in today's dollars), the Kindle sold out in less than six hours and would go on to move 240,000 in its first year.[6] But his thinking definitely favored the multipurpose device over one devoted to the sole function of book reading, and Apple had been developing and patenting elements of a touchscreen and stylus-based tablet. However, after intense internal debate it was decided to prioritize development of the iPhone over what was to become the iPad.

The iPhone was introduced in 2007, delaying introduction of the iPad until 2010, giving Amazon three years to demonstrate decisively that Americans were far from illiterate: That year, Bloomberg News estimated that 10 million Kindles would be sold.[7] How Jobs found a way to have his cake (denigrate reading) and eat it, too (sell lots of print and e-books), we shall see.

The third contender in the e-reader race was Barnes & Noble. Given its 726 stores in 2008, the chain was a potential El Dorado of e-reader sales.[8] Unfortunately, their initiative got off on the wrong foot because of a patent dispute that lasted years and postponed release, until November 2009, of its dedicated e-reader Nook (a name that provoked some controversy).[9]

The arena was set for the entrance of the gladiators.

12. GAME ON

(2007-12)

Behold the Sony, Nook and Kindle
Spawn of Gutenberg and Tyndale.
Every day a new device
Bids to win a market slice.
Tipping-pointward e-books tramp,
Overrunning print-books' camp.
Ten years ago a callow stripling,
Now every month shows volume tripling.

N OVEMBER 19, 2007, WAS THE BIRTH DATE of the Kindle, a day I had dreamed of with messianic anticipation since that moment, some twenty-two years earlier, when I flashed on the notion of a portable e-book reader. Though Jeff Bezos's device was not perfect, as can readily be demonstrated by subsequent fixes, refinements, upgrades and generations (a total of eleven, at this writing), the designers and engineers had thought of everything. At that moment in technological history, the Kindle was state of the art, the fulfillment of untold thousands of hopes—and millions in investments.

Amazon's press release on launch day was artfully designed to answer buyers' questions, quell their anxieties, address their skepticism, satisfy their curiosity and appeal to their budget:

> Nov. 19, 2007—Amazon.com (NASDAQ:AMZN) today introduced
> Amazon Kindle, a revolutionary portable reader that wirelessly down-
> loads books, blogs, magazines and newspapers to a crisp, high-

resolution electronic paper display that looks and reads like real paper, even in bright sunlight. More than 90,000 books are now available in the Kindle Store, including 101 of 112 current New York Times Best Sellers and New Releases, which are $9.99, unless marked otherwise. Kindle is available starting today for $399 at http://amazon.com/kindle.

The release went on to sound every bell and whistle known to technology:

- Downloads Content Wirelessly, No PC Required, No Hunting for Wi-Fi Hot Spots
- Books can be downloaded in less than a minute and magazines, newspapers, and blogs are delivered to subscribers automatically. Amazon pays for the wireless connectivity for Kindle so there are no monthly wireless bills, data plans or service commitments for customers.
- Reads Like Paper. It reflects light like ordinary paper and uses no backlight, eliminating the eyestrain and glare associated with other electronic displays such as computer monitors or PDA screens.
- The Kindle Store currently offers more than 90,000 books, as well as hundreds of newspapers, magazines and blogs. Customers can search, browse, buy, and download from this wide selection wirelessly from their Kindle. The same Amazon shopping experience customers are accustomed to is offered in the Kindle Store, including customer reviews, personalized recommendations, 1-Click purchasing, and everyday low prices. Additionally, Kindle customers can download and read the first chapter of most Kindle books for free.
- At 10.3 ounces, Kindle is lighter and thinner than a typical paperback and fits easily in one hand, yet its built-in memory stores more than 200 titles and hundreds more with an optional SD memory card. Additionally, a copy of every book purchased is backed up online on Amazon.com so that customers have the option to make room for new titles on their Kindle knowing that Amazon.com is storing their personal library of purchased content. When customers order a Kindle, it arrives from Amazon.com ready to use. There is no software to load or set up. Customers are immediately ready to shop, purchase, download and read from Kindle.

- Amazon is adding new book, periodical, and blog titles to the Kindle Store every day.
- Publishers and authors can submit their content and make it available to Kindle customers by using Amazon's new Digital Text Platform (DTP), a fast and easy self-publishing tool that lets anyone upload and sell their books in the Kindle Store. Customers can take their personal documents with them on their Kindle.
- Built-In Dictionary and Wikipedia
- Long Battery Life
- Kindle has a standard-layout keyboard that makes it possible for users to search the Kindle Store, their entire library of purchased content, and Wikipedia.org
- Annotation and Bookmarks
- Ergonomic Design
- Adjustable Text Size

The announcement swept consumers away; Kindle was a runaway hit, reportedly selling out in under six hours.[1]

A number of factors contributed to the buying frenzy. For one thing, the public was not just ready for the Kindle, they were hungry for it, even at a price of $399 (the basic model lists for $109.99 at this writing). A generation of users was now completely at ease accessing media on desktops, laptops, smartphones and PDAs, and many of them had already been reading books, newspapers and magazines on them. Manipulating this sleek utilitarian device to read books was second nature, and even if you had never used a computer, it was a piece of cake. And the appeal was built into the brand: Amazon, a company built on customer satisfaction, was a respected, if not beloved, consumer brand, particularly when it came to books.

With the exception of the Sony E-Reader, Kindle had the field to itself, and would have it for two years.

Perhaps none of those reasons weighed as heavily, however, as the low price of the e-books themselves: $9.99. In a post on the Kindle Forum years later, as reported by Dennis Abrams on the Publishing Perspectives website, the company justified the price on the grounds that their device was competing against "mobile games, television, movies, Facebook, blogs, free news sites, and more."

However valid those reasons may sound, many critics assailed the gambit as predatory, and it became a flashpoint for a series of battles—waged with publishers, competitors, and even authors themselves—that was to climax in all-out legal war.

How was Amazon able to price their e-books so low and make a profit? The answer is, they weren't. In order to conquer their competitors, they maintained their low price even if it meant losing money. The strategy of "loss leader" was a tried-and-true, if nasty, way to drive competitors out of the water. But if one has enough money to sustain losses, it can be a winner. If it was good enough for nineteenth-century robber barons, it was good enough for Jeff Bezos. The plain fact was, Amazon had unlimited money to burn, and Bezos had not invested years of time and tons of money only to lose a price war.

In his memoir, *Turning Pages*, former CEO of Macmillan John Sargent describes his dismay about the practice of loss-leading after a disturbing conversation with Bezos. "You buy the book for ten bucks," Sargent reasoned, "then you sell it for ten bucks, and meanwhile you are paying people and rent, building warehouses and investing in technology. That cannot work."

Bezos smiled and said, "Trust me, it works."

"OK," conceded Sargent, "assuming it works, how long can you keep doing it?"

"Forever," Bezos replied.[2]

When I was notified of Amazon's policy, I became alarmed, fearing that if they were cutting their prices, they were also going to cut royalties to make up for their losses. I pounced on their contract and was relieved to confirm that our percentage—50 percent at the time—was guaranteed and would remain intact. And though we would be collecting less money per sale, the low price generated a higher volume of sales, just as the Amazon brain trust had shrewdly predicted.

Amazon's game plan worked beyond expectations: After selling out on launch day, the device was not restocked until the following April, five months after release.[3] According to Tidbits.com, "In 2009 Amazon controlled 90 percent of the e-book market." Within three years, Forrester Research's James McQuivey reported, Amazon had sold some 4 million Kindles in its various versions.[4] In 2014 alone, $5 billion worth of Kindles were sold.[5]

It took a year for Amazon to absorb our E-Reads titles into its Kindle inventory. Our 2008 gross from all retailers was $180,000. In 2009, however, it climbed to $306,000; and in 2010 we began to feast on Amazon's bounty, collecting $724,000, of which $269,000 was Kindle revenue. (These numbers do not include Amazon print and e-book sales through Ingram's LightningSource program.)

Our good fortune was tinged with anxiety when Michael Gaudet, who had steered our boat through many a shoal, announced he was moving on. (As of mid 2025, he is Vice President of Digital Operations at Hachette Book Group.) Gaudet graciously agreed to stay on to train a replacement. By that time, our operation was large enough to require a helmsman with extensive technical and publishing experience. We placed an ad on craigslist. Among the respondents was a personable young man named Anthony Damasco, who had strong programming, marketing and website design skills. On the downside, he confessed he had no publishing experience whatsoever. "I didn't even know what rights were," he recently told me with a laugh, "but I was used to taking on customers without knowing anything about their business." Like other businesses he had worked for, he saw E-Reads in terms of system management. He was a quick study, and after a few of my long-winded tutorials, he grasped where we fit into the book business and set his sights on moving us to the next level.

With his gung-ho leadership abilities, Anthony took charge of our team, which was now made up of specialists in editorial and production, cover design and contract management. He focused on building an internal communications system that enabled everybody to harmonize their efforts and pinpoint delays in the assembly line. He also focused on perfecting our royalty system, cleverly employing a different highlight color for each vendor, enabling me to easily track revenue ebb and flow.

Our offices occupied the second floor of a brownstone on East 74th Street on the Upper East Side of Manhattan. At the south end were the E-Reads rooms, at the north end, the agency's. E-Reads' precincts were eerily quiet, the staff's labors performed on mouse and keyboard as opposed to the sound, at the other end of the building, of me yakking on the phone, the ambient clamor of twentieth-century publishing before it succumbed to the digital hush of business telecommunications in the new era.

We met weekly in my office, where I debriefed their achievements and set goals, quotas and strategies. I have never been a tough boss, but the one thing I demanded was clear communication. Only if technical issues were explained to me in plain English was I able to make cogent decisions. My rule was "When in doubt, don't." I sometimes glimpsed my crew exchanging side eyes when I requested what must have sounded like a childishly simplistic explanation, and I took some teasing. But I didn't mind: A good-natured ribbing was preferable to committing a costly error because I didn't understand the implications of a decision. A measure of success was that I was able to explain complex processes clearly in my blogs and newsletters to authors.

In 2009, two years after the launch of the Kindle, a serious rival materialized in the form of Barnes & Noble's Nook, whose release, as mentioned, had been delayed by a patent hassle. But B&N dramatically made up for lost time.

The Nook boasted a small color LCD touchscreen to navigate the gadget's menu, but otherwise, in its E Ink display and most other features, it wasn't wildly different from the Kindle or the Sony E-Reader. However, its $259 price undercut the Kindle by $140, which certainly contributed to the Nook's initial success. As Michael Kozlowski explained in *Good e-Reader*, "[Barnes & Noble's] relationship with major publishers through its bookstore business ensured that B&N could gain access to their digital wares and more books were added to its online portfolio." B&N quickly stocked the Nook with over one million titles in the first year and "within a relatively short time," Julie Bosman wrote in *The New York Times*, "had picked up 25 percent of the e-book market, slicing into Amazon's dominance."

A year later, the company released the Nook Color, a full-color handheld with a 7-inch screen that Kozlowski describes as "a tablet billed as an e-reader." In addition to books, it utilized apps that provided multimedia, drawing a clearcut line between dedicated e-book readers and a full-media experience, a line that would soon separate the digital world into warring parties. Consumers ate up B&N's device, buying millions of them for the holidays—one million on Christmas Day, 2010, alone![6] Some savvy book editors found a way to adapt their Nook to a tablet for reading manuscripts.

Barnes & Noble gave Amazon stiff competition: sales of the Nook soared from $105.44 million in 2010 to $695.18M in 2011 to just a bit shy of one billion dollars in 2012. Revenue from the device fattened E-Reads' coffers too, climbing from $58,000 in 2010 to $130,000 in 2011 to nearly $250,000 in 2012.

All this activity generated intense speculation about what Apple was up to. Though cofounder Steve Jobs shrouded his company's efforts in secrecy (including code names for projects in development) and affected disdain for his competitors, he could not have been unaware of the impressive sales numbers being posted by Amazon and Barnes & Noble. But from the very beginning of his reign, he had set his sights on a device that carried other media besides books. As early as 1983, in a speech at the International Design Conference in Aspen, Jobs said:

12. GAME ON

> What we want to do is we want to put an incredibly great computer in a book that you can carry around with you that you can learn how to use in 20 minutes. That's what we want to do, and we want to do it this decade. And we really want to do it with a radio link in it, so you don't have to hook up to anything. You're in communication with all these larger databases and other computers.[7]

Jobs's vision was not to be realized for two more decades, but he never abandoned his idea, assembling the device in bits and pieces, trials and errors, starts and stops. As early as 1991, Apple designed a stylus-operated tablet, and in 1993, the company released a personal digital assistant called the Newton MessagePad. It failed to win consumer hearts, and Jobs pulled the plug in 1998, at a head-spinning loss of $100 million.[8]

In October 2001, Apple released the iPod, a portable media player principally employed as a music device, but some versions carried other media like games, videos, photographs and even—though requiring no small measure of user manipulation—books.[9] Two years later, Macintosh introduced the iTunes Store, which became the library and vendor for iPod's music. The device was a runaway bestseller: By 2007, it had sold 100 million units, and when it was finally discontinued in 2022, the total was 450 million, dwarfing the sales performance of dedicated e-book readers.[10]

Sometime in 2004, it was reported that Apple had filed design patents for a touchscreen-driven tablet. The iPad was taking shape. Yet surprisingly it was relegated to second place in Jobs's priorities. He felt that another gadget should take precedence, this one called the iPhone. He had a point: In 2007, iPhone's first year, Apple sold 1.39 million of them. The next year it sold 11.63 million, grossing almost $7 billion. From there, sales growth was almost exponential through 2012 and continues to fly in the hundreds of billions of dollars to this day.[11] In 2018, Apple became the first U.S. company to be valued at $1 trillion.

After some internal debate as to whether Apple should allow third-party developers to produce Apple-compatible apps, the company released a software development kit in March 2008, and the following July, the App Store was launched, licensing outsiders to create applications for Apple devices as long as they met the company's strict standards and agreed to pay a 30 percent commission to Apple. Among the candidates was Amazon, which seized an opportunity to piggyback on its rival. In March 2009, Amazon released a Kindle app for iPhone and iPod Touch, enabling users of those devices to download content from the Amazon store. Thus, the two companies became "frenemies," using each other to make money on e-books. How much money Amazon netted might be wondered, because the commission paid to use the Apple store must have cut sharply into Amazon's e-book profits.

The tense business relationship between Apple and Amazon lasted a couple of years and would deteriorate into enmity over several issues, particularly e-book pricing. The resulting clash was to convulse the publishing industry, as we shall soon recount.

Meanwhile, Apple's work continued apace on the multimedia tablet, and at last, on January 27, 2010, the iPad was announced. By March, Apple was taking preorders. With a 9.7-inch screen it was tablet in size and navigability, offering 256 megabytes of random access memory and as much as 64 gigabytes of storage. Appleinsider.com writes:

> The tablet's signature achievement was bringing the iPhone's multi-touch interface to a much larger display, allowing it to behave more like a laptop. Though it still lacked an open filesystem or much customization, Apple developed a custom version of the iPhone OS [operating system] for it, for instance letting people use apps and the homescreen in any orientation—unlike the iPhone, which at the time was strictly vertical.[12]

The first generation iPad was heavy and clunky, with a $1,024 \times 768$-pixel screen that was sharp and brilliant but LCD-backlit, causing glare and draining the battery. Eyestrain was not a problem with the Kindle's gentle grayscale and E Ink display, which used minimal juice. The contrast in recharge requirements between the iPad and the Kindle was dramatic: every ten hours for the former, every ten *weeks* for the latter.

On April 3, 2010, the iPad was launched domestically, selling 300,000 off the bat and 1 million within a month. Sales reported for the second half of 2010 were 7.46 million, according to Statista.[13] In 2011, Apple moved 32.9 million units. That year, too, Apple released a second generation iPad that was 33 percent lighter, 15 percent thinner and generally more powerful.[14] Thereafter, new generations were announced regularly, featuring different designs, sizes, power and accessories. Sales continued in the tens of millions of units annually, attaining a total of 425 million as of March 2021.[15]

12. GAME ON

Given the astronomical sales of the iPod music player, it's easy to understand why Jobs did not give top priority to the creation of an e-book reader. The iPod had demonstrated that music outsold books by a ridiculous multiple. But he did not neglect literature entirely. Mac devotees were reading e-books on their iPhone. And now, simultaneously with release of the iPad in 2010, Apple announced iBooks, located in the Macintosh iTunes App Store. Users could import the e-reader app from iTunes free of charge. However, the powers that be realized that pre-installing the iBooks app in the device would increase ease of access. This was done for a subsequent version of the iPad, and the tactic did indeed provide a big boost to Apple's book sales.

At launch, the iBooks library offered 60,000 volumes.[16] (It isn't clear whether that included 30,000 public-domain titles accumulated by Project Gutenberg and offered free of charge.) Though Apple didn't break out iBook sales in its financial reports, a 2015 article in *Publishers Weekly* suggested the app had performed quite satisfactorily, to say the least. "Not only has the Apple iBooks store recently recorded 1 billion e-book downloads since its launch," iBooks store director Keith Moerer said, "[but] the online e-book seller is also attracting 1 million new users a week and sells e-books in 51 countries."[17]

I wasted no time offering the E-Reads list to Apple. But there was a big hurdle: We would have to become a developer and furnish our content as an app, a technical challenge far beyond our skill set. That was out of the question. Hell, the only thing we'd done for other vendors was to furnish digital files, and now I was supposed to create E-Reads the App?

Happily, good old LightningSource came to our rescue, offering to become our agent, in effect, converting our books to the Apple format and taking care of all the niggling details. They were already doing that for our Amazon content. Every month they issued Amazon royalties to us, not just for U.S. Kindle and print sales, but for UK, Canada, Germany, France, Spain, Italy, Brazil, Japan and India, as well. Currency conversion calculations were a pain in the ass, but we lived with that vexation, because Lightning took care of an even bigger pain, foreign taxes (some countries levy them on revenue going out of their country). Even domestic sales were fraught, because a number of U.S. states charged taxes on book sales, and we'd have had to deal with 50 tax-table headaches. Lightning took care of everything.

In 2011, the first year that we grossed over $1 million, iPad royalties began flowing abundantly, totaling $50,931.

That same year, a new entry appeared on our balance sheet. We started paying advances. Originally, when we launched E-Reads, we had decided against this practice. We made one exception: In 2008, we licensed 100 titles from Kensington, for which we did pay advances. Steven Zacharius, the company's President and CEO, said, "Until E-Reads, nobody had ever paid advances for e-rights."

But that was the only instance until 2011. Adding advances to production costs at the outset of our operation would have put us out of business. Instead, as mentioned, we charged a production fee to authors, but rather than ask them to pay it up front, we laid it out on their behalf and recouped it from royalties. As they were getting 50 percent royalty in exchange, they accepted the tradeoff as fair, and it enabled us to manage our finances prudently. Our business model was working. Unfortunately, it did not appeal to literary agents.

Advances are so deeply embedded in the reptilian brains of literary agents that the notion of a publisher acquiring a book for no money upfront is tantamount to an attack with a butcher knife. Another anathema for agents was charging authors for production expenses. The easiest way for an agent to get fired is to tell a client, "I've made a terrific deal for you: No money upfront, plus you're going to pay to have your book produced."

These agents, however, represented authors we wanted for our list. What could we do to win them over?

Our dilemma was exacerbated by the entrance of a number of rival e-publishers. Thanks to progress in technology, it was no longer that hard or expensive to start an e-publishing company. Once you had a digital text and access to stock photos for your covers, the tools for producing e-books and PODs were cheap and reliable. I began to hear the footsteps of competitors close behind us. Some of them offered more than our 50 percent royalty; others dropped charging authors for production expenses. We did have advantages: a ten-year head start, a tried-and-true business model, an experienced technical and production team, and a powerhouse list of books and authors. Still, we were concerned.

One of these rivals was Byron Preiss, a creative and imaginative publisher, packager and author, a charming man and personal friend (my son Charles worked for him in the summer of 2003 and fall of 2004). He was endlessly curious, inventive, enterprising and uncannily plugged in to the future, working in areas like multimedia and graphic fiction long before they entered the mainstream. Television luminary Carl Reiner, with whom he had worked, said he was thirty years ahead of his time.[18]

Byron had packaged a number of properties in CD-ROM but now set out to produce them as e-books. He launched ibooks in 1999, the same year as E-Reads and with the same objective. His website's mission statement could have been our own: "ibooks™ harnesses the latest digital technology to get its books published at the speed of the 21st century." Byron's list was not as broad or viable as our own, but he was catching up.

One afternoon in July 2005, Harlan Ellison called my wife and me with devastating news. A moment later Charles called, sobbing; he had heard the same news: Byron Preiss had been killed in a traffic accident at the age of 52.

Byron was a man of countless ideas and gifts. *Publishers Weekly* called him "a visionary with an unbreakable love for books and unmatched energy for new ideas."[19] Although his life was cut short before ibooks could find its footing, had he been spared: I'm certain he would have found a way to make ibooks a dominant enterprise.

Byron's company's assets were acquired by another firm, and in time, and after a long legal wrangle, the ibooks trademark was acquired by Apple, which capitalized the *b* to make it iBooks.[20]

Another competitor, Open Road, was particularly formidable. Founded in 2009 under the leadership of the flamboyant former HarperCollins CEO Jane Friedman, it was backed by millions of dollars in equity financing, more in one tranche than E-Reads had earned in its entire history.[21] Suddenly big investors smelled money in e-books (at least in Jane's e-books).

Friedman and her team pitched Open Road's virtues to major literary agents representing those heavy-hitting authors who still had scruples about the legitimacy of e-book publication. Surprisingly, there were still a lot of them. Many were dubious about a format that was so remote from conventional books. In addition to its seductive message of wealth, Open Road promised these holdouts a "secret sauce," namely publicity and promotion, benefits that few independent e-book publishers, including ourselves, could afford to offer authors.

I greeted Jane, an old friend, with the opening line of Walt Whitman's "Song of the Open Road," "Afoot and light-hearted I take to the open road." I hoped that her company's addition to the rising tide would lift E-Reads' boat, but I was also very nervous that this behemoth could swamp us. In time Open Road would play a transformative role in the life of E-Reads, but in 2011, they were our adversary, and we urgently sought a competitive edge.

The answer was to offer advances. No other e-book publisher was doing it, not even Open Road, with all its riches. (Nor am I aware of any independent e-book publisher doing it today.) The good news was that we could now afford not only to pay advances but also to phase out our practice of charging authors for production expenses. After all, advances are simply a form of paying authors with their own money; you just have to be careful not to overpay. Thus, as an inducement to agents, our advances made them look good to their clients. This shift in our business model gave us a definite competitive advantage, and we attracted a number of agents who had been hanging fire.

By far the biggest advance we ever paid was to Harlan Ellison, the brilliant but curmudgeonly fantasist. You'll recall that back in the 1980s he had become a client of my agency. His temperament was why the term "PITA factor" was devised; PITA stood for "Pain In The Ass." I had created a tongue-in-cheek formula in which you divided client commissions by the degree of aggravation they inflicted. The lower the yield, the higher the PITA factor, and Harlan's was almost off the charts. His rants were nuclear, but they throbbed with wit and brilliant insight. You are excused for 3½ minutes to watch his immortal "Pay the Writer" diatribe on YouTube.[22]

Despite his periodic eruptions of fury, however, we were fast friends, an achievement I managed by not taking his fulminations too seriously, however lengthy and venomous. One day he said to me, "I love you because you treat me like I'm your only client." Without breaking a smile, I replied, "You *are* my only client." I think he half believed it.

He was no respecter of my privacy and frequently opened a phone call with "I realize it's Yom Kippur . . ." or "I realize it's New Year's Eve. . . ."[*] One day he prefaced a call with "I realize it's Easter Sunday. . . ." He had discovered a typo in his just-published book and wanted me to call the publisher (in church! on Easter Sunday!) and demand that he destroy the edition. How I teased him down from that tantrum should qualify me for canonization.

Ellison was not just skeptical about e-books but downright hostile to them, and though almost all of his books were out of print, I could not persuade him to put them into the E-Reads program. But one day in 2008, he called. "I'm up the creek. We're broke, and I'm gonna lose my house."

The house was his beloved Ellison Wonderland. This fabled edifice was a literally fantastic four-thousand-square-foot residence in Sherman Oaks, California, a mini-museum filled with autographed first editions of science fiction

---

[*] Agents could be PITAs too. I'm told of one who demanded to speak to a certain editor who was in a late stage of labor.

classics and adorned with art, including original paintings of his covers. A climate-controlled room contained first editions of comic books. And all about the house were dazzling collectibles—first editions of his own and his friends' books, and toys, games, tchotchkes from the worlds of sci-fi, movie and television, collected over decades of a stellar, colorful and stormy career.

I saw my opportunity and offered him advances of one thousand dollars per book for print and e-book rights to thirty-two titles. Thirty-two thousand dollars was a big nut for our humble little company, but this was a once-in-a-lifetime opportunity. Ellison not only restrained his antagonism to the digital medium but pitched into refreshing the books with his customary zeal, editing each one for the launch of a matched paperback set that we called the Harlan Ellison Collection. Though it wasn't as prestigious as, say, the collected works of Thomas Carlyle, I was very proud of it, and I'm happy to say it didn't take long for us to recover our investment. (You can see Harlan on YouTube proudly opening six cartons filled with copies of the Harlan Ellison Collection, with his handsome face on the cover of each volume.)[23] If you figure he earned on average $2.00 royalty per sale (either print or e-book), our thousand-dollar advance for each book would earn out after 500 units were sold, and he would then begin collecting royalties.

Around this time a windfall from across the Atlantic fattened our bank account. Although our fantasy and science fiction e-books were selling well in the United States, they were moving feebly in the United Kingdom owing to the fact that the Brits had been slow to board the e-book bandwagon. I concluded that maybe our e-books would sell better if they were issued by a native British house rather than imported from the U.S. I reached out to Orion, one of England's leading science fiction publishers, and offered them UK rights to our sci-fi, fantasy and horror titles. They had been looking for a wedge to pry open their home market, and our deal comprised more than six hundred e-book titles. Orion paid us an advance against royalties for each title plus handsome fees for our production files.

Our bottom line was starting to look very good, and with so many mergers and acquisitions in the news, I began to indulge in speculation about what I could get for E-Reads in the marketplace. I put out some feelers, but it must have been obvious that I was ambivalent about putting my baby up for sale, as no one expressed serious interest.

Until Amazon.

In May 2009, Amazon had launched a print book publishing division. Their first couple of years were dedicated to reissues of out-of-print books, translations and some works by self-published stars. But two years later they created a couple of imprints in the romance and crime genres. After engaging a major publishing executive, Laurence Kirshbaum, former CEO of Time Warner Book Group, the company created a third imprint called 47 North, specializing in fantasy, science fiction and horror. Larry, who had been a literary agent before his stint at Time Warner, was a personal friend, and after he took the Amazon job, I reached out to him to test the waters about buying our company.

Any fisherman who has cast a light line into the sea only to hook a huge thrashing tarpon will understand how I felt when Larry told me Amazon was interested.

But then, why would they not be? By 2011, we had somewhere around two thousand titles, many of which were fantasy, science fiction and horror by branded practitioners of those trades. Our list was a perfect fit for Amazon's. The 9 billion–ton gorilla wanted to embrace me! But then I got to thinking: If I delivered my list to them, it was likely they would demand exclusivity. That is, they would not distribute our titles to any of the other retailers we did business with—Barnes & Noble, Apple, Google (which had started an e-book store in 2009) and many more. A multiplicity of vendors was our hallmark. I would have to abandon all those vendors for an exclusive relationship with Amazon. I suspected that many of our authors—uneasy with if not downright hostile to Amazon—would not be thrilled to be in the Amazon family.

I don't know if I merit a badge of courage or a dunce cap, but I walked away from Amazon. Subsequently, someone told me I was legendary for having said no to Amazon.[24]

As things turned out, it was a wise decision. Amazon's entry into the ranks of publishers provoked bitter enmity among booksellers, a community that had been decimated by Amazon's ruthless competitive tactics. Barnes & Noble, the giant's biggest rival, refused to carry Amazon's list of print editions, causing the colossus to resort to a subterfuge. In 2012, it made a deal with Houghton Mifflin, a distinguished old-line Boston publisher boasting a nineteenth-century pedigree. Houghton Mifflin agreed to sell an imprint by the name of New Harvest, which just happened to be almost completely composed of Amazon titles. Surely, Amazon reasoned, retailers would not pass up the list of so esteemed a publisher as Houghton Mifflin. But, led by Barnes &

Noble, the retailers closed their loading bays to New Harvest. Amazon by any other name was still Amazon.[25]

Smarting under this rejection, and aware how important it was for authors to see their print books in brick-and-mortar stores, Amazon tried another gambit. In 2015, it opened a bookstore in Seattle, and after testing the reactions of customers (and authors), it opened another sixty-seven across the country.

The scheme flopped.

Some observers noted that the stores were not warm and welcoming, nor did their staffs really understand the tastes of their clientele and community the way homegrown local bookshops do. There was more to selling books than the algorithms (like "customers like/bought") that worked so well for the online operation. It was also speculated that Amazon had done too good a job of conditioning consumers to prefer online shopping to physical stores. According to Jeffrey Dastin of Reuters, the stores' revenues "often failed to keep pace with growth in the retailer's other businesses." Eventually (early in 2022), Amazon announced it was closing all of its physical shops.[26]

Apple's Store, on the other hand, not only succeeded but had a profound impact on civilization's progress into the digital era. A July 10, 2018, post by Stephen Silver in the *Apple Insider*, celebrating the App Store's tenth anniversary, stated it succinctly:

> The App Store has not only grown exponentially in its ten years of existence, but it's also been at the forefront of all sorts of innovations in technology, culture and entertainment over the course of the decade. The App Store has helped facilitate major growth in the content streaming revolution, as well as geolocation, e-commerce and even online dating, while also forever changing what it means to be a software developer.[27]

Our biggest year was 2012, grossing over $1.5 million. We were hitting on all cylinders, with money coming in from Amazon, Barnes & Noble, Apple, Baen, Baker & Taylor, Diesel, Fictionwise, Google and Kobo. The biggest contributor by light years was Amazon, which, according to researcher Shehraj Singh, owned a 72 percent share of the e-reader market as of 2023, followed by Barnes & Noble with 10 percent. "Alternative apps," Singh points out, "make up the remaining 18 percent."[28] Sales of the iPad, in other words, were in the "also ran" column of your racing form. That certainly jibes with our own stats. For example, in the third quarter of 2012, Apple brought in $23,821.00 for E-Reads and Barnes & Noble took in $64,729.00, but Amazon generated a whopping $298,317.00, more than three times the combined total of its competitors. Nevertheless, iPad revenues contributed to the heady realization that our company had arrived.

It would be great to end the e-book story at this glorious moment in time, but despite the robust growth and prosperity of the industry over the coming decade, with annual unit sales in the hundreds of millions, ugly weather was darkening the landscape.[29] Publishers and bookstore chains were starting to bristle under Amazon's hegemony.

Since its founding thirteen years earlier, Amazon had been the object of polar ambivalence among publishers. A 2012 posting on Publishing Perspectives by its then Editor in Chief, Edward Nowatka, put it this way:

> The company's very name inspires equally vitriolic doses of love/hate/angst in such a wide variety of constituencies that it's impossible to keep track. Publishers fear it but are dependent on it. Readers adore its low prices and convenience, but hear from some circles that they should boycott the company. Writers know they need it — to sell their books. And self-publishers realize that without Amazon, the whole self-publishing phenomenon might never have taken off in the first place.

The website's caption under the photo of Amazon founder Jeff Bezos said it all: *Jeff Bezos: Savior, Satan, or something in-between?*[30]

As the first decade of the twenty-first century progressed, publishers' resentment rose higher and higher as the retailer's chokehold on the industry tightened, and they looked for ways to fight back. One defensive measure they essayed was direct-to-consumer selling, in which publishers marketed their print books directly to the public without the intervention of a retailer (especially one named Amazon). This had worked for Doubleday when it was both a publisher and a bookshop chain, but that had ended in 1990 when Barnes & Noble/B. Dalton acquired the chain. Selling print books directly to the public required a great deal more marketing skill than most publishers—with the exception of some genre houses like sci-fi specialist Tor Books and romance behemoth Harlequin—had. And the whole idea was self-defeating anyway, simply because it meant that publishers selling their books at list price would be competing with their own retailers, who were able to heavily discount the same books.

With the introduction of Amazon Books in 2009, it was clear that Amazon was becoming less of a partner and more of a menace, driving publishing companies and bookstores into a corner—and sometimes leaving them there. Publishers that ignored digital options or relinquished them to Amazon ended up irrecoverably stranded on the wrong end of the paradigm shift. Retailers that remained stuck in brick and mortar and hard goods were heading for a big fall.

It came in 2011, when the Borders bookstore chain filed for bankruptcy, closing some 674 stores and eliminating 11,000 jobs.[31]

The underlying causes were complex, but the shift from analog to digital was at the heart of it. "Borders," explained Yuki Noguchi in an NPR podcast, "went heavy into CD music sales and DVD, just as the industry was going digital. . . . It expanded its physical plant, refurbished its stores and outsourced its online sales operation to Amazon." Noguchi quotes Morningstar investment researcher Peter Wahlstrom, who said, "In our view that was more like handing the keys over to a direct competitor."

Even consumers aided in hastening the chain's demise, thanks to an insidious practice called "showrooming," whereby customers browsed merchandise at brick-and-mortar stores, selected what they wanted, then went home to purchase it online, where it was cheaper—in all likelihood from Amazon.[32] This was not a passing fad but a consumer phenomenon that reached epidemic proportions. *USA Today*, in a 2013 article entitled "Strategies: 10 ways you can combat showrooming," reported that "more than a quarter of shoppers with smartphones checked out prices online while in a store."[33]

"The physical bookstore," bankruptcy law professor Laura Bartell was quoted as saying in the *Detroit Free Press*, "has become a thing of the past."[34]

This was not quite true. Enterprise abhors a vacuum, and into the void left by Borders sprang up a host of independent bookstores. The gigantic Barnes & Noble bookstore chain, facing the same market forces as Borders, was weakened but rescued itself in large measure because of a greater commitment to digital technology: specifically, its e-tail website, bn.com, and its Nook e-book reader. And it didn't hurt that B&N owned the Sterling

Publishing Company, a house with some five thousand nonfiction titles.[*] (Sterling was acquired by Hachette in 2024.)

The death of Borders sent a magnitude 8.0 shockwave throughout the publishing world. Aside from the immediate damage—the evaporation of millions of dollars owed to publishers (a portion of which was payable to authors)—the long-term effects were devastating. The loss of acres of shelf space reduced projected print runs, making publishers more blockbuster-driven than ever before. As one editor put it, "Every book has got to be a home run." The chain's collapse arguably altered the very nature of our literature and reading tastes, for Borders was a leading carrier of genre fiction, a staunch category that represented about 25 percent of trade book revenue.

The handwriting was on the wall: If things continued in the direction they were going, Amazon had the power to crush the publishing industry. Jeff Bezos's operation brought in more money in six months than the entire trade book industry made in a year.[35] "They could buy the whole damn publishing industry with their spare change," one editor told me.

Something had to be done about Amazon.

---

[*] Barnes & Noble continued to struggle until it was sold in 2019 and restored to health under its CEO James Daunt.

13. WHAT WERE THEY THINKING?

(2013)

Scarcely had the year unfurled
When iPad swept into the world.
Compensation posed by Apple
Proved bewildering to grapple.
Big Six lawyers scratched their noddle
Parsing Jobs's business model.
Fretting that his precious Kindle
Market share would sharply dwindle,
Bezos took extreme exception,
Spoiled his e-book foe's reception,
Took his ire out on Macmillan,
Cast it in the role of villain,
Fired a broadside at John Sargent
Cost Macmillan beaucoup d'argent.

F OR ALL OF ITS MARVELS AND WONDERS, the App Store was the locus of one of the darkest episodes in the history of e-books, a monumental price war that that sucked Apple and five major publishers into its vortex, climaxing in a highly publicized, embarrassing and costly lawsuit. The details are complex, but in essence, the conflict stemmed from two issues, both provoked by Amazon's introduction of the Kindle.

The first was the timing of releases. Amazon, with the support of many customers, insisted that publication of e-book editions must be simultaneous with that of hardcovers. This suspension of accepted logic was a major departure. Until now the traditional sequence for most fiction and nonfiction was publication of the hardcover, followed, after nine or twelve months or longer, by a cheaper reprint edition of the book, either a trade paperback (large size) or mass market (pocket size). This gave rise to the catchphrase, "I'll wait for it to come out in paperback."

The practice of staggering the release of versions of a work is called "windowing"—that is, creating an exclusive *window* for the most viable or valuable version, to maximize its sales potential before releasing other versions. In movies, it might mean theatrical release first, followed by DVD or streaming or television exhibition. In publishing, windowing generally refers to withholding release of cheap reprints of a book until the higher-priced edition has run its course.

Drawing on this tradition, publishers equated e-books with paperback reprints: You published the hardcover, gave it its exclusive sale window, then when sales diminished, you would bring it out as an e-book. You could release the e-book at the same time as the paperback; after the paperback; or even instead of the paperback. *But not at the same time as the hardcover!*

Delaying e-book release made perfect sense from the publishers' viewpoint, for they feared that the cheap would cannibalize the dear. Many of them were perturbed and even alarmed by this prospect of the simultaneity Amazon was demanding. To what degree the practice vitiates print book sales today, we will never know, but I personally believe it must be substantial.

In addition, a new menace had risen up: piracy. The sudden availability of digital texts converted from printed books created a treasure chest for hackers, for whom extracting the files from websites was child's play. Pirates were starting to steal the book business blind—a 2019 Forbes.com posting estimated the losses at $300 million annually.[1] Though windowing e-book release would not cure piracy, it would at least enable publishers to maximize print volume sales before succumbing to the inevitable shoplifting of their digital texts when the e-book came out.

In July 2009, Dominique Raccah, Publisher of Sourcebooks, a formerly independent publisher (now owned by Penguin Random House), issued a stout defense of windowing. "Formats have windows," she told Kassia Krozser on the Booksquare website.

> We know when we [book publishers] put out different formats in the
> lifecycle of a book. So we shouldn't be releasing e-books at the same
> time that we release a hardcover book. We should be releasing e-
> books when we release the trade paper or mass market of the hard-
> cover and can then price appropriately to that. To me the decision is
> analogous to a new release in movie theatres; we don't expect that
> movie to be immediately available on DVD.[2]

From the viewpoint of many Kindle owners, however, windowing was an
outrage. A very passionate contingent howled in opposition, asserting that they
had paid a lot of money for their e-reader and expected immediate access to
every just-released book, publishers be damned. In fact, some of them devised
a ploy to punish publishers who withheld their e-book reprint editions: In a tactic known as "review bombing," they ganged up to downgrade certain new Amazon releases to one star, the lowest you can get, potentially depressing the value
of the work.

Hannah Johnson, writing on Publishing Perspectives, described what happened to one book.

> The recently released book, *Game Change* by John Heilemann and
> Mark Halperin (published by Harper), was well-received by newspa-
> per and online book reviewers for its examination of the 2008 presi-
> dential campaign in the USA and how Obama won the election. So
> why does this book have [a 2 out of 5] star-rating on Amazon.com?[3]

Steven W. Beattie, reporting on this action in a *Quill & Quire* posting,
counted 113 one-star ratings for this book out of 193 reviews. Some of those
who had assigned one star out of pique were completely blunt about their reasons. One wrote:

> I purchased a kindle to have immediate access to books as soon as
> they are released, not stand in a virtual line waiting for the publisher
> to get with the program. Had Amazon informed me before I bought
> a kindle that new releases would be available only once interest in
> the hardcover version had died down completely, I wouldn't have
> purchased a kindle.

Another:

So the publishers decided to make Kindle users wait for six weeks to buy the hot "new" political tome? In six weeks, we'll be on to something else. Sorry, I spent a premium on my Kindle. I'm not going to run out and buy another hardback for $25.

A third:

I know some think that rating a book as 1-star due to an e-book release delay is unethical or inappropriate. But I think these publishers need to get the message that it's not the right thing to do, and they need to work out their differences with Amazon. The best way to do that is to do this rating and damage the overall average—hopefully that hurts them where it hurts (the pocketbook).[4]

Terrified of Amazon and its militant minions, the Kindle users, publishers had no choice but to swallow the bitter pill and abandon windowing. Eventually, the simultaneous release of "P" and "E" (as print and electronic editions are known in publishingese) became the rule, but at that moment in time, the issue was still being debated. It would rear up again and come to a head in a prodigious showdown.

The second cause of friction was the low price, $9.99, that Amazon charged for many new e-book releases and bestsellers. Depending on which side of the sale you were on, the price was a gift (consumers) or a curse (publishers, who pointed to the havoc wreaked to the music industry by iTunes pricing digital singles at $.99 each). Those publishers may well have been thinking about Tower Records. Capitalized at $1 billion in 1999, Tower filed for bankruptcy in 2006, thanks in no small measure to the shift in recordings from CDs to digital downloads (and to massive piracy).[5]

Combined with the elimination of windowing, Amazon's bargain basement pricing structure for e-books posed cruel dilemmas for trade book publishers, particularly the major houses, which numbered six at that time: Macmillan, Penguin, Hachette, Simon & Schuster, Random House and HarperCollins. Amazon was a third-party licensee and therefore free to set whatever price on Kindle editions it wished. The price it wished was $9.99, an amount that competitors were hard put to match and publishers even harder put to make money on.

Amazon justified this figure by claiming it could move almost twice as many copies (1.74 times, to be precise) of an e-book priced at $9.99 than it could one listing at $14.99. On this point the Authors Guild expressed skepticism. Novelist Roxana Robinson, President of the Guild, stated in Huffington Post that "they don't include any source for the study that they cite, so it's unclear how they reached the numbers they put forth."[6]

Not only were profit margins on Kindle sales painfully thin, but publishers also feared that cheap e-books would seriously eat into sales of their hardcovers. After all, if print and e-book editions were published simultaneously, why wouldn't consumers choose the cheaper version? Yes, many book lovers preferred a tangible object to hold in their hands and display on their bookshelves. But many others were fine with an evanescent text that vanished when they were through reading it.

There was also an important psychological component at play: Publishers worried that consumers would start to believe that *all* books should be priced at $9.99, a misperception that could depress the market for hardcover books, which at this time were priced at $24.99 on the average. Choosing between $24.99 and $9.99 was a no-brainer.

Amazon, whose policies almost invariably sided with consumers, was not moved by these or any other arguments: $9.99 was the e-book shopper's sweet spot. Amazon had publishers over a barrel, and things looked pretty dire. "If readers started to believe the right price for an author's bestselling new work was less than ten bucks," Macmillan's John Sargent relates in his memoir, "many industry executives feared the ecosystem of publishing would collapse."

The table was set for an intrepid scheme.

According to the transcript of the subsequent lawsuit, it happens that "on a fairly regular basis, roughly once a quarter, the CEOs of the Publishers held dinners in the private dining rooms of New York restaurants, without counsel or assistants present, in order to discuss the common challenges they faced, including most prominently Amazon's pricing policies." At one such event, a publishing executive promised his boss that "he would raise with his competitors their options to confront the "potentially dominant role played by . . . Amazon . . . in order to control their strategy and pricing." This person was quoted as saying, "I hate [Amazon's] bullying behavior and will be happy to support a strategy that restricts their plans for world domination."[7]

The only path out of this dilemma was to find a way to raise the price of e-books. Or to be precise, to make *Amazon* raise the price of e-books.

The launch of the Apple Store in 2008 cast a ray of hope over these grim circumstances. The store was to serve as a retail agency for vendors, collecting a 30 percent commission on their products and services. Suddenly publishers saw a way out of their bind, a means of regaining control of e-book pricing. This came to be known as the "agency model": Publishers could utilize the Apple Store as their agent to sell e-books for whatever prices they saw fit. The e-books would list between $12.99 to $14.99, prices that would make a consumer think twice before selecting an e-book over a hardcover. Even after Apple's 30 percent vigorish, the publishers would net a nice profit on e-book sales.

Of course, there was a little hitch to this course: Who would buy an e-book for $14.99 or even $12.99 when they could get it from Amazon for $9.99? Indeed, consumers could buy one for $9.99 from Apple itself by using the Kindle app in the Apple store!

There had to be a way to force Amazon to raise its e-book prices. And there was. If publishers discontinued, or threatened to discontinue, furnishing e-books to Amazon and to shift their business exclusively to Apple's agency model, wouldn't that work? Maybe. But the only way for this stratagem to succeed was for the publishers to coordinate their plans and act in concert.

Which was a brilliant solution except for one thing—it was against the law.

Surely, their common sense (or legal counsel) would have made them aware that the plot they contemplated was the formation of a combination in restraint of trade in contravention of the Sherman Antitrust Act. Yet, in some mystifying act of mass myopia, they managed to subdue their qualms. In December 2009, they stepped over the line separating independent from collective action. Knowing that Apple would be releasing its iPad and opening its iBookstore in the next month, executives at major houses began discussing—with Apple and each other—a variety of deal points, including higher list prices for e-books. As Denise Cote, the judge in the ensuing lawsuit, stated in her formal opinion, they had formed a "joint venture." Indeed, one executive even referred to the group as "the Club." What made matters even more conspiratorial was that their action had the active encouragement and participation of Apple itself, right up to Steve Jobs, who was certainly no babe in the woods when it came to matters of antitrust law.

Leading Apple's initiative was Eddy Cue, Senior Vice President of Internet Software and Services. Early in December 2009, "Cue's team contacted the Publishers to set up meetings the following week to discuss an 'extremely confidential' subject," Judge Cote explains. "Apple's requests for meetings in New York were an exciting turn of events for the publishers, and prompted a flurry of telephone calls among them."[8] There was also a flurry in the contracts departments of these publishers: They were drafting agency-model contracts with Apple that contained many provisions almost identical to each other. A key term was a Most Favored Nation–like clause, which in effect guaranteed that no publisher could get a more advantageous deal from Apple than the others. Indeed, it "also imposed a severe financial penalty upon the [publishers] if they did not force Amazon and other retailers similarly to change their business models and cede control over e-book pricing to the Publishers."

December and January saw a proliferation of activity as the parties got in line—all except for Random House, which opted out of the scheme:

> The CEOs of the [publishers] made over 100 telephone calls to one another in the short period of time between December 8, when Cue first contacted them, and January 26, when the Agreements were signed. In the critical negotiation period, over the three days between January 19 and 21 . . . [executives of the five publishers, Penguin, HarperCollins, Hachette, Simon & Schuster and Macmillan] called one another 34 times, with 27 calls exchanged on January 21 alone.

By the end of the week of January 18, the transcript states, "four of the five Publisher Defendants had put Amazon on notice that they were joining forces with Apple and would be altering their relationship with Amazon in order to take control of the retail price of e-books. It was clear to Amazon that it was facing a united front."

When a last-minute case of cold feet developed at HarperCollins, Cue suggested that Jobs call James Murdoch (Rupert's younger son) of News Corp, HarperCollins's parent company, and "tell him we have 3 signed so there is no leap of faith here." The holdout finally signed. "Thus" wrote Judge Cote,

> . . . in less than two months, Apple had signed agency contracts with five of the six Publishers, and those Publisher Defendants had agreed with each other and Apple to solve the "Amazon issue" and eliminate retail price competition for e-books. The Publisher Defendants would move as one, first to force Amazon to relinquish control of pricing, and then, when the iBookstore went live, to raise the retail prices for

e-book versions of New Releases and NYT Bestsellers to the caps set by Apple.

By mid-January 2010, it was clear to Amazon that the publishers had linked arms. The time had come for the publishers to make it official. But who was going to bell the cat? Macmillan CEO John Sargent felt he should meet face-to-face with Amazon's brass. Judge Cote's account is colorful:

> Skipping the Launch to which he had been invited, Sargent flew instead to Seattle, accompanied by [Brian Napack, President of Macmillan]. Thus, Macmillan, the smallest of the five Publishers, did the honorable thing and delivered its message in person. Sargent did not expect the meeting to go well. As he put it, he was "on [his] way to Seattle to get [his] ass kicked by Amazon."

Sargent was right. When he confronted Amazon, demanding it switch to the agency model, a furious Jeff Bezos lashed out by ordering the Amazon Buy buttons on Macmillan's website to be turned off, sending tremors of terror throughout the book industry—including authors whose books were hostages to this conflict.[9]

But the other four publishers came to Macmillan's aid, informing Amazon that they, too, were committed to Apple and the agency model. As a result, in the words of DearReader.com, "Amazon learned that five of the six publishers agreed to the Agency model and that these five accounted for about half of Amazon's ebook business and thus Amazon caved to Macmillan."[10]

The publishers' united front accomplished its mission to break Amazon's hammerlock on the e-book business and got Bezos to turn Macmillan's Buy buttons back on. "Over the weekend," Judge Cote writes, "it became obvious to Amazon that its strategy had failed." The forces arrayed against Amazon—determined publishers, angry authors, upset customers, and even nervous investors (Amazon stock had dropped almost 8 percent in January)—were too formidable to oppose any longer.[11] On January 31, Amazon announced that it would "capitulate and accept" Macmillan's agency terms, "because Macmillan has a monopoly over their own titles, and we will want to offer them to you even at prices we believe are needlessly high for e-books."

Needless or not, the result of these synchronized actions was an immediate jump in the prices of Kindle's e-books. The new ones averaged 18.6 percent higher.

But the publishers' triumph over Amazon was hollow. "Shortly thereafter," states the judge, "Amazon sent a letter to the Federal Trade Commission complaining about the simultaneous nature of the demands for agency from the Publishers who had signed with Apple." It took two years for facts to be gathered and arguments marshaled, but finally, in April 2012 the U.S. Department of Justice, joined by 33 states (plus Canada and the European Union), initiated civil antitrust suits against Apple and the five publishers.

The publishers settled out of court, at a cost of $160 million.[12]

Macmillan's Sargent was the final holdout, but the pressure on him was overwhelming. In his account, Sargent tells how Judge Kimba Wood, mediating settlement negotiations, delivered a message "from the government of the United States. 'They told me to ask if you fully realize they are capable of bankrupting your company,' " she told Sargent. He had no choice, and he writes, "The cost of the case for us . . . was close to $30 million."

Apple, however, contested the complaint. Their case—*United States v. Apple Inc.*—presently proceeded to trial. Apple's defense was alternately ingenious and disingenuous. For example, Cue argued that the use of the phrase "solves Amazon issue" in his communications

> referred to pricing e-books *in the iBookstore above $9.99 and was not a reference to raising prices across the industry or eliminating Amazon's ability to set prices.* Indeed, Cue protested at trial that, throughout its negotiations with the Publisher Defendants, Apple was concerned only with the pricing that would prevail in the iBookstore and sought *only to "fix" Amazon's pricing or "solve the Amazon issue" in its own e-bookstore.* [italics mine]

The drama of the battle portrayed in Judge Cote's opinion makes for entertaining reading, and a visit to 952 F. Supp. 2d 638 (S.D.N.Y. 2013) will reward your trouble. But Apple's arguments were to no avail. On July 10, 2013, Judge Cote announced her decision:

> Based on the trial record, and for the reasons stated herein, this Court finds by a preponderance of the evidence that Apple conspired to

restrain trade in violation of Section 1 of the Sherman Act and relevant state statutes to the extent those laws are congruent with Section
1. A scheduling order will follow regarding the Plaintiffs' request for
injunctive relief and damages.

Her final proclamation, "SO ORDERED," rang like a hammer on an anvil:
Apple appealed the decision all the way up to the U.S. Supreme Court, but the
Court declined to hear it, and Apple ended up paying a fearful settlement of
$450 million.[13]

We look back at this episode and shake our heads in bewilderment. *What
the hell were they thinking?* Surely these sophisticated executives understood the
risks of transgressing antitrust laws. What suspended their sound judgment?

The answer, I believe, can be understood only by placing ourselves in the
publishers' shoes at that moment in time. Amazon's virtual takeover of the retail
book business had already created immense turmoil, anxiety and financial hardship for publishers. The introduction of the Kindle should have been their sal-
vation, but Amazon's onerous economic model, releasing ultracheap e-books
simultaneously with print editions, only added to the publishers' stress. The
Kindle was the only game in town; Amazon could pretty much write its own
ticket—and did.

The revenue from those $9.99 Kindle sales after Amazon took its cut and
authors were paid their share did not compensate publishers for profits lost on
hardcover sales. We can only excuse, or at least explain, their abandonment of
caution by concluding that they were desperate, and teaming up to fight Amazon was the only way to save their companies—and perhaps the industry—from
financial catastrophe. One can conjecture that if Apple had been content to let
things unfold without pushing publishers, and if the publishers had refrained
from coordinating with each other, everything would have been copacetic. But
the record of the proceedings strongly suggests that without those factors, and
with the vast black shadow of Amazon blanketing the book business, nervous
publishers might well have chickened out of the Apple deal.

The court record confirms the publishers' level of distress: "As early as
February 2009, Apple's Cue recognized that '[t]he book publishers *would do almost anything* for us to get into the e-book business.' " [italics mine]

Apple's motives seem easier to understand: profit, profit, profit. With sales
of millions of iPads projected, the opening of the App store with its free iBooks
app, five out of the Big Six publishers in Apple's pocket, and e-book prices in the

$12.99 to $14.99 range, Apple envisioned minting a fortune in the e-book space over which Amazon had held near total hegemony just a short while before.*

And how do we judge Amazon's role? It's easy to cast them as the villains in this drama. Their arrogant, predatory and intransigent policies provoked the conflict, and undoubtedly, they profited from the increases in e-book prices. But in their favor, it can be argued that they resisted those price increases for the sake of their customers. Former Amazon engineer Gregg Zehr reminds us that Jeff Bezos often said, "Start with the customer and work backwards." Bezos had started with the customer and, working backwards, arrived at a $9.99 price and elimination of windowing. That this policy might hurt his vendors was secondary on his value spectrum.[14]

And however harshly we may judge Amazon's behavior, they were literally and legally victims of a conspiracy. As Judge Cote asserted, Amazon was not on trial.

> This trial has not been the occasion to decide whether Amazon's choice to sell NYT Bestsellers or other New Releases as loss leaders was an unfair trade practice or in any other way a violation of law. If it was, however, the remedy for illegal conduct is a complaint lodged with the proper law enforcement offices or a civil suit or both. *Another company's alleged violation of antitrust laws is not an excuse for engaging in your own violations of law.* [italics emphatically mine]

In other words, two wrongs don't make a right. Amazon's wrong was not made right by Apple's. Apple paid dearly for its wrong. Amazon walked away unpunished.

Yet, the wheels of justice did not grind to a complete halt. It took ten years for the remedy suggested by Judge Cote—"a complaint lodged with the proper law enforcement offices"—to be applied to Amazon's business practices. But on September 27, 2023, *Publishers Weekly* announced that

> the Federal Trade Commission, supported by 17 state attorneys general, finally filed its long-awaited antitrust lawsuit against Amazon

---

* Their projections were accurate: close to 8 million iPads sold the first year alone.

yesterday. In a 172-page complaint, the government alleged that the e-tailer "uses a set of interlocking anticompetitive and unfair strategies to illegally maintain its monopoly power." The use of that power allows Amazon "to stop rivals and sellers from lowering prices, degrade quality for shoppers, overcharge sellers, stifle innovation, and prevent rivals from fairly competing against Amazon."[15]

Amazon-aggrieved Dennis Johnson, co-founder of Melville House, summed it up: "About fucking time."

14. DISPLACED PERSONS

(2007-11)

Self-pubbed authors by the score
Enlisted others in their corps,
Advocating insurrection
From Establishment subjection.
Alas, they learned success depends
On sales to family and friends.
The ancient proverb's still conclusive:
The wealth of Indies is elusive.

D ESPITE THE COMMONLY HELD IMAGE of glamour, influence and knife-to-the-throat power, the daily life of most literary agents is absorbed with routine management of small matters punctuated only occasionally by thrilling negotiations. And those are fewer and further between than is commonly believed. No, the daily life of a literary agent is usually occupied by the most mundane of recordkeeping, scorekeeping and housekeeping chores of the kind that any dental receptionist or shoe store manager would recognize. As Leslie once quipped, "When you lie down with agents, you wake up with paper clips and rubber bands."

I mention this, because I was so immersed in the day-to-day preoccupations of running both an agency and a publishing company that I was slow to recognize signs that authorship was drifting towards a crisis.

A key factor was the collapse of the marketplace from hundreds of publishers to a mere handful. The explosion of chain superstores and the demands they made on publishers for More! Better! Faster! had driven countless under-financed houses into the arms of a few well-heeled suitors. And though the acquired publishers ostensibly continued their lives as imprints of the houses that acquired them, many of them eventually lost their founders, their staff, their identity, their spirit and, ultimately, their *raison d'être*.

The consolidation of the nineties had generated intense pressure on publishers to justify acquisitions, causing them to raise the profit bar to levels unachievable by authors who had previously made money for their publishers and a comfortable living for themselves. Improved metrics and the increased intimacy between publishers and chain store buyers enabled publishers to predict a book's performance and, if its projected sales fell on the wrong side of the midlist penumbra, to decrease their commitment or reject it entirely. In acquisition meetings, the minimum print run for a hardcover or mass market paperback might now be pegged at double or triple or an even greater multiple of the previous threshold. Sales that worked at 10,000 copies before a publisher was acquired had to work at 50,000 afterwards. Authors—even established ones—whose projections did not attain those levels, were not invited to return. The phrase "death spiral" was applied to authors whose print orders were halved with each new book until they plummeted below the level of profitability, the publishing equivalent of what in baseball is called "the Mendoza line," a dismal .200 batting average that will send players down to the minor leagues and maybe out of the game entirely.

The term midlist did not originally have a negative connotation; it simply described books of general interest written by competent and reliable authors for a small but viable readership. However, in this blockbuster atmosphere the word came to connote failure. For instance, the jacket copy for *How Not to Write a Novel: Confessions of a Midlist Author* by David Armstrong, published in 2003, reads:

> Being a published novelist is not always as glamorous as it seems from the outside. There are the depressing, ill-attended readings, the bitchy writers' conventions, and the bookshops that have never heard of you and don't stock your book. All of these will be familiar to any writer who, like Armstrong, falls into the category euphemistically known in publishing as "mid-list." The reality is that for every J. K. Rowling there are 1,000 David Armstrongs; for every writer who is put up in a five-star hotel and flies first class courtesy of their

publisher there are 1,000 who sleep on friends' floors during book tours and dine at highway service stations.

Even a rock band got its two cents in: A group called the Decemberists produced a song entitled "Midlist Author" with the refrain,

> You're never the best but you're never the worst
> Why even bother? (even bother?)
> You'll never be last but you'll never be first.[1]

To evade this perdition, authors and publishers devised a number of strategies. Some, resigned to exclusion from major houses, turned to small, underfinanced independent presses. Victims of the Death Spiral wrote new works under pennames so that their agents could present them as gifted debut authors with no track records whatsoever. Others—those who could afford to work without pay for six or twelve months—wrote on spec that big juicy commercial novel that they had vowed to produce someday. They couldn't afford to do it, but neither could they afford further decline of their careers. And so, they put their savings at risk for this shot at the gold ring. Some made it; most did not.

Clearly, there were too many authors and too few publishers. For a growing number of professional writers, quiet desperation became career panic. Everything seemed to be hitting them at once: vanishing markets, pressure to self-promote, an unfamiliar new medium and a growingly unrecognizable publishing landscape. The odds against acceptance by big publishers had become lottery-sized. "Slush piles" were inundated. ("Slush" is book industry lingo for manuscripts submitted to publishers by unrepresented writers.) When the nation was younger, trade houses were open to "over the transom" submissions, and a number of such works ended up getting published.[*] However, towards the end of the twentieth century, Big Publishing abandoned the practice and shut the doors to unagented authors.

To understand why, you only have to do the math. Up to the end of the twentieth century, trade book houses were inundated with thousands of

---

[*] Wiktionary explains that over the transom "refers to the idea of a writer tossing a manuscript through the open window over the door of the publisher's office." One summer, as Leslie and I were leaving a conference, a writer tossed a manuscript through the window of our car, narrowly missing her face.

unsolicited submissions. My best guess is no fewer than five thousand annually for the bigger companies. The scripts were usually assigned to freelance or junior editors to read. If an editor did nothing else but read slush, they might process four manuscripts a day (one doesn't have to read every page of a bad book) or roughly a thousand per year. Thus, you needed at least five editors to read those five thousand submissions. The starting salary of junior editors was about $17,000 in the early 1980s—about $50,000 today, so the cost for five editors to read slush was about $85,000; today that would be more than a quarter of a million dollars.[2] We are talking about a substantial investment to find a needle in a haystack.

The cost of slush processing might be recouped if these editors discovered three or four publishable books a year, or even one bestseller. But the odds of that happening were minuscule, and the exceptions only served to prove the rule. *Ordinary People* by Judith Guest was plucked out of the unsolicited pile at Viking Press and went on to become a very big book and an even bigger movie. But according to *The New York Times*, it was also the first such manuscript accepted by Viking in *twenty-seven years*! A 2010 article in *The Wall Street Journal* reports that in 1991, a book by Mary Cahill, *Carpool*, was pulled out of the slush pile by a Random House editor and went on to become a bestseller. "That," said reporter Katherine Rosman, "was the last time Random House, the largest publisher in the U.S., remembers publishing anything found in a slush pile."[3]

At length, towards the end of the twentieth century, publishers ceased to believe they could find publishable books in those bushels of unsolicited submissions and concluded that it was more cost-effective to stop accepting over-the-transom manuscripts and spend the savings on more productive appropriations (like lunches). They informed authors that unsolicited manuscripts would no longer be considered, and submissions must come from agents.

Incidentally, it wasn't just slush that was dumped on agents. Numerous tasks that had once been the exclusive province of publishers fell upon our shoulders, too. Like many of my colleagues, I found myself composing jacket copy, critiquing cover art, editing manuscripts, creating websites and fashioning marketing campaigns. I made it a policy to be copied on all emails to and from my clients, and I never hesitated to interject my two cents into an editorial dialogue when I felt it would help the author, the publisher or both. I developed expertise in disciplines far afield from the services I performed at the outset of my career.

Though most agents were (and still are) willing to take a chance on an intriguing pitch by a newcomer, agents were soon flooded and compelled to turn away all but a handful of queries, leaving countless authors *sans* publisher and *sans* agent. To make things worse, the number of trade book publishers had been severely reduced to a handful during the great consolidation of the 1980s and '90s. In my own experience, out of every hundred queries I received, I might request one manuscript, and out of every hundred manuscripts *requested*, I might take on one. That's 10,000 to 1 odds that your query ended up with me representing your book. I would guess those numbers were the same for many of my colleagues. One, Kate McKean, posted in her newsletter in October 2024, that she was some five thousand behind in unread queries.[4]

Where could authors turn? In the analog era gone by, they might have resorted to so-called vanity publishers, glorified printers who published any book that an author was affluent enough to subsidize. Vanity publication cost thousands of dollars, and the investment was seldom recovered, for distribution was limited to a few outlets plus friends and family.

The new century offered better tools for writers who were contemplating self-publication, but this option was not much more desirable than the previous one. It still bore the smell and stigma of vanity publishing, and subsidizing publication of their work was simply against their religion. For another thing, there was not yet an effective means for a self-published author to reach a broad audience—or any audience at all. The tools for marketing and distribution were still embryonic.

There were inspiring exceptions, however. Richard Bolles's multimillion-selling career guide series *What Color Is Your Parachute?* inspired others to try it. To accommodate them, a number of companies like Xlibris, 1st Books and iUniverse were launched in the late 1990s, offering print-on-demand services using PDF text files. Aside from listing the books' availability on retail websites, the marketing and promotion performed by these publishers was minimal. "What Xlibris doesn't do," novelist and bookseller Susan Taylor Chehak wrote in *Publishers Weekly*, describing her self-published nonfiction book *Don Quixote Meets the Mob*, "is edit, publicize or market any of the books it prints. The author is on her own." The process was not cheap: *PW* reported the cost at somewhere between $459.00 and $1,900.00, equivalent to $740–$3,050 at this writing.[5]

By 2004, the three leading self-pub presses had produced some 45,000 books. That may seem like a lot, but it was a fraction of the number of unsolicited manuscripts. The good news was that as the decade progressed, new tools became available for publishing one's own books cheaply and efficiently and delivering them to a widening audience.

When Amazon released the Kindle in November 2007, it also made available an e-book application called Digital Text Platform, designed to allow authors to upload and sell their own e-books in the Kindle store.

Although e-book publication was good enough for some authors, however, many felt that the only "real" book was a printed one. Amazon had a solution for that, too. A South Carolina company called BookSurge had developed a print-on-demand program to help independent authors produce and publish their own print books.[6]

Amazon already had a setup for self-published authors called CreateSpace, but as failory.com pointed out, CreateSpace only "focused on the supply of tools for self-publishers," whereas the South Carolina outfit "maintained a catalog of thousands of different titles that were available for on-demand printing."

BookSurge came to the attention of Amazon, which acquired it in 2005. The news item about the acquisition was a quiet little squib in some trade publications, but for me, it clanged like a firehouse bell. I had resumed writing a column around this time, a blog called Publishing in the 21st Century. In a posting called "Gone Today, Gone Tomorrow," I noted that financial analysts had underappreciated Amazon's acquisition of the e-book and POD companies. "But because both deals carry the promise of virtual distribution of books," I wrote, "the implications of those deals are profound."

> BookSurge's print-on-demand capacity holds the potential for a virtual distribution model in which the hard copy of a book does not exist until a customer orders it, nor is it printed until the customer pays for it. It's hard to imagine a model that is simpler or more economical. Surely Amazon CEO Jeff Bezos must have asked himself why he still owns nearly 4 million square feet of warehouse space, no small portion of which is devoted to storing books, and employs 7,500 people to process them, when he could simply forward orders to a printer who will manufacture and drop-ship the copies directly to customers.

In December 2009, Amazon merged the services of BookSurge with those of CreateSpace.[7] Thus, Kindle Direct Publishing (KDP) was born.

I said that the implications of these deals were profound. For one thing, the avenue for unpublished (and in some cases unpublishable) writers was now a highway. KDP (and its rivals) offered, to tyros and professional authors alike, a marketplace for their books at completely affordable prices, with step-by-step customer service to facilitate development from draft to release. By 2017, when Amazon folded CreateSpace into its Media on Demand program, it had published 1,416,384 titles.[8]

Self-publication exposed a fatal flaw in the conventional publishing process, and some very savvy authors were about to make a lot of money exploiting it.

The flaw was the one I had been harping on for thirty years: the practice of consignment distribution. The returnability of books was a hole in the dike of publishing revenue. Publishers paid not only for shipping their books *to* stores but for shipping the unsold copies *back* to their warehouses. Thus, when a book doesn't sell, they lose twice.

Publishers considered the consignment model to be an immutable cost of doing business. Such was their resignation that even bestsellers were actually considered successful if their return rate did not exceed 25 percent! A May 2016 *Publishers Weekly* article actually exulted over the presumably sunny news that "in 2015 compared to 2014, the trade paperback return rate from reporting publishers was the lowest, around 20 percent, while hardcover returns were 26 percent and the mass market return rate was 48 percent."[9]

To add to the burden of returns, the overhead of traditional publishers—office rent, staff salaries and benefits, printing and distribution, to say nothing of the unrecouped portion of their advance to authors—was enormous, driving undercapitalized publishers into the arms of the giant houses or out of business altogether.

The boom of the '80s and '90s had disguised these fault lines, but as the new century progressed, the wreckage of that binge became sharply visible. The Great Recession of 2008 hit publishing hard, causing layoffs and declines in revenue.[10] Novella Carpenter, writing on SFGate, noted that "one publishing behemoth, HarperCollins, lost 75 percent of its operating income during the first six months of 2008. Overall, the publishing industry has struggled as bookstore sales—and the economy—have slowed drastically."[11]

Among the most precious resources damaged in the upheaval were authors. Casting their eyes about, they saw a bleak marketplace. The number of viable trade book publishers had shrunk to a handful of giants preoccupied with courting high-profile writers with dazzling offers. The mass market paperback, once a fertile breeding ground for young talent, had upgraded into an elite reprint medium for hardcover blockbusters by superstars. Neophyte storytellers were forced to cast their fates to small presses, many of which had modest finances, limited distribution and meager marketing means. The system of building authors from break-even to breakout had all but given way to one that demanded they hit the ground running with polished masterworks. Where and how were writers supposed to develop their skills without the support of editors and the feedback of readers?

A cadre of smart professional writers figured it out. As the decade progressed, a solution began to emerge. Authors were learning how to utilize desktop publishing tools to write, edit and format their books, select typefaces, design covers and upload the package to the Internet. They learned to build handsome and dynamic websites and employ a growing suite of social media ornaments to enhance their platforms. Self-publication was becoming a viable option. All it needed in order to blossom was a vehicle to distribute widely to readers.

And now, with Amazon's KDP platform, they had one.

Courageously, they walked away from the lures and emoluments of the establishment and opted for the speed, simplicity and austere economics of self-publication. These authors leaped, and for many of them, as it is said, the net appeared. In January 2023, Wordsrated, a research data and analytics group, reported that one thousand authors—commonly called "indies" (short for "independents")—made over $100,000 from their self-published books in 2022.[12] The "wealth of the Indies," indeed!

As I said earlier, stripped to its essence, the book publishing process can be defined as a writer, a reader and a server. The new technology disintermediated the complexities between writer and reader, reducing the path to publication to a few cheap and easy steps. Authors could establish a one-on-one relationship with their audiences and learn firsthand how fans responded to their books. Utilizing KDP as their server and applying a variety of desktop editorial, graphics and design tools, authors like J. A. Konrath, Barry Eisler, John Locke, Hugh Howey and Amanda Hocking demonstrated how self-publication could be mined to achieve astonishing fame and fortune on Amazon's 70 percent royalty. Another company, Smashwords, founded in 2008, distributed a vast library of self-published books and paid royalties comparable to Amazon.

14. DISPLACED PERSONS

In numerous blogs and blasts, Konrath became the voice of self-publication. He also articulated the case against traditional publishing in a host of incisive blogs. He could afford to bite Big Publishing's hand, because it no longer fed him. Some examples:

> Legacy publishers* are a cartel. I suppose it could be a coincidence that the Big 6 all have exactly the same (low) royalty structure, and shockingly similar contract terms. But collusion seems easier to believe, and this collusion is aimed at limiting the income and power of authors. Legacy publishing contracts are painfully one-sided.
>
> Legacy publishers have zero transparency when it comes to things like sales, returns, print runs, and inventory, and keep authors in the dark.
>
> Legacy publishers fix prices. That's what the agency model is. Even worse, these prices are too high and hurt authors' sales.
>
> Legacy publishers sometimes fail to edit.
>
> Legacy publishers abandon books, releasing them into the market without any push at all.
>
> Legacy publishers pay royalties twice a year. Are you freaking kidding me?!? It's 2012! Why are their accounting and payroll departments stuck in 1943?
>
> Legacy publishers embraced returns for full credit. This is the biggest fail in the history of retail, and the reserves against returns practice has screwed thousands of authors. Isn't it funny how whenever you hear about an author auditing a publisher, unreported sales are always discovered?[13]

In March 2011, thriller novelist Barry Eisler electrified the book industry with the announcement he had turned down a two-book $500,000 offer from a traditional publisher in favor of self-publication. "I just couldn't go back to working with a legacy publisher," he explained. "The book is nearly done, but it wouldn't have been made available until spring of 2012. I can publish it myself a year earlier. That's a whole year of actual sales I would have had to give up." His astounding decision inspired a dialogue with Konrath that reads like a political manifesto. Its thesis: Traditional print publishing is antiquated, unprofitable

---

* "Legacy" is the same as traditional.

and poorly managed, and must eventually succumb to the countless advantages of self-publication.[14]

Their interchange ranged from history—

> JOE: As a self-publisher, you can get your books to readers much faster—often by a year or more—than a legacy publisher can. You don't have to deal with the ungodly amount of time we've both spent touring, booksigning, and travelling. There's no wasting time or getting frustrated with publishers' mistakes. You're in complete control of your own career, whereas before you were at the mercy of a corporation that treated you like just another product—a product that it wasn't very good at selling in the first place.

to business—

> JOE: We figured out that the 25 percent royalty on ebooks they offer is actually 14.9 percent to the writer after everyone gets their cut. 14.9 percent on a price the publisher sets. [Note: After Amazon's 30 percent cut, 70 percent of the retail price goes to the publisher, which pays 25 percent to the author's agent, who takes a 15 percent commission leaving 14.875 percent. payable to the author.]

to politics—

> BARRY: If for generations you've been the lord of the land worked by your peasants, and you suddenly find yourself needing the peasants more than they need you, if you find them making new demands you don't have the negotiating leverage to resist, you'll probably find yourself resentful because damn it, this just isn't the way God ordered the universe!

and even to philosophy—

> JOE: Those who don't study history are doomed to repeat it. I also think the Upton Sinclair quote is appropriate: "It is difficult to get a man to understand something, when his salary depends upon his not understanding it." Denial is a powerful opiate.

14. DISPLACED PERSONS

The E-Book Revolution had spawned E-Book Revolutionaries! Even I joined their ranks when I began vociferously advocating the overthrow of the consignment system of book distribution and replacing it with a completely print-on-demand model. I regarded POD, with zero returns, as the panacea for the evils about which I had been berating publishers for decades. Naive me! Publishers' addiction to consignment was incurable. They needed huge printings to "blow out" their big books. Even if half of the copies came back from these blitzes, the low unit printing cost per copy made it profitable for major publishers.

Acknowledging that the industry was trapped in the consignment business model, I proposed one last remedy to the industry, which, with unabashed vanity, I dubbed the Curtis Plan. It was predicated on the assumption that 50 percent of copies of any given book will be returned to the publisher. This means authors will earn royalties on only 50 percent of the distributed books. What's worse, their 50 percent will be parceled out to them a little bit at a time, until years after publication, their book's sales are finalized.

Under the Curtis Plan:

1. Publishers would pay royalties on copies *distributed*.
2. The total royalty would be paid within 30 days of printing.
3. Authors' royalty percentage would be 50 percent of the royalty paid in the traditional model.

By way of example (and you may skip this passage if you are not a nerd), let's examine a traditional deal. Suppose your publisher paid you a $10,000 advance against a 10 percent royalty; based on the $25.00 list price of your book, that's $2.50 per book sold. And further suppose your publisher distributed 50,000 copies. Your possible royalty would be $115,000—$125,000 less the $10,000 advance your publisher recoups. Does your publisher pay you $115,000? No, because they must hold most of it against the possibility that 50 percent of copies will be returned. Your net sale will therefore be 25,000 copies. When your publisher determines that returns are final, the balance of your royalties will be released to you. Not counting your advance, you will have earned $52,500 but it will have taken you five years or longer to collect it. Because of the complexities of tallying sales and returns and held reserves and released reserves in any given accounting period, your publisher must retain a large royalty accounting department.

Now look at your deal under the Curtis Plan. The same $10,000 advance, the same 50,000 copies distributed. But instead of 10 percent royalty based on copies *sold*, you get 5 percent royalty on copies *distributed*. Your net royalty is exactly the same $52,500 as the traditional deal. But it is *paid to you in full within 30 days of distribution of your book*. Adopting this model, publishers can shut down their royalty departments, since these transactions can now be calculated by one person (or that person's child) with pencil and paper.

As quixotic as this scheme may sound, the Curtis Plan was based on the simple and elegant model used by Fawcett Paperbacks until it was acquired by CBS in 1977. When I proposed it to some publishers, they said it was great—but it was too hard to reverse consignment, even though it no longer applied to modern conditions.

So much for visionaries!

Every revolution produces collateral damage, and though "E-Book Revolution" sounds harmless enough, it actually took a serious human and social toll. Publishers and agents, those stalwart gatekeepers of author careers, were now judged to be superannuated, stuck in the analog mud.

Like Barry Eisler, a number of authors severed the connection with their publishers, some of whom had supported them from their very first book. But what could publishers offer that would tempt them back? After all, the standard cut of e-book revenue offered by publishers was 25 percent (and still is, at this writing), compared to the 70 percent or more that authors got by directly distributing their books with Amazon, Apple, Smashwords, Barnes & Noble and other e-tailers.

Trade publishers tried to woo back departing authors with the argument that traditional print publication was far more prestigious than print on demand. Those arguments often fell on sympathetic ears, for even though their income was the envy of their peers, many self-published authors were ambivalent about completely parting company with legacy publishers and were willing to sacrifice profit for prestige. Indeed, when legacy publishers started throwing big advances at them, some authors realized that traditional print was profitable, too. They therefore made hybrid deals, giving publishers print rights but keeping e-rights for themselves.

One such bargain was made by John Locke, acclaimed the first e-book writer to sell a million copies (and who chronicled the process in *How I Sold 1 Million eBooks in Five Months*).[15] He agreed to let his publisher, Simon &

Schuster, retain the sales and distribution of his printed books, but he kept e-book revenue for himself.[16] Hugh Howey made a similar hybrid deal.[17]

Even the great curmudgeon himself, Joe Konrath, eventually linked up with a legacy publisher. In March 2018, Tainted Archive blogger Gary M. Dobbs wrote:

> It will seem a strange move to his fans, but self-publishing superstar Joe Konrath, who has spent many years crusading against traditional publishers, calling them unnecessary in the modern digital world, has signed with Kensington Books in a bid to bring out his bestselling eBooks in mass market paperback. Konrath's blog, A Newbies Guide to Publishing, contains countless posts in which the author hits out at traditional publishing, Legacy publishing as Konrath calls it, and yet the author has now signed with Kensington to bring his back list to the mass market.[18]

Publishers weren't the only entities to be ditched. Many literary agents were let go—literally disintermediated—by clients who realized that their representatives were unable to perform meaningful services for self-published authors. A few agents did manage to hop on the technological bandwagon and perform commissionable duties for their clients. Literary agent Steve Axelrod had foreseen the lucrative potential of indie writers and was instrumental in bringing some of them into the mainstream publishing culture. One of his stars was Amanda Hocking, who had created a mammoth online following for her original and sexy paranormal e-books in Amazon's KDP program. Axelrod took her to St. Martin's Press and signed her to a conventional print/e-book deal for $2 million. He subsequently sold one of her previous self-published trilogies to St. Martin's as well.[19]

But too many other agents were slow to adapt or couldn't adapt at all, and not a few authors left them and went off on their own.

Fortunately, my dual identities as both agent and publisher gave me an advantage. As an agent, I continued to handle traditional deals for our authors, and as a publisher, I reissued their out-of-print books. For works I was unable to sell as an agent, I simply published them in E-Reads, thus keeping client defections to a minimum.

Though the list of self-published authors who hit the jackpot is impressive, their numbers were dwarfed by the horde of writers—estimated at two to three million—who succumbed to the siren song of easy publication and sought a stake in the e-gold rush. Jason Ward, in *The Writing Cooperative*, wondered whether self-publishing was "the new slush pile."[20] These aspiring writers may not have realized that aside from their storytelling gifts, successful self-published authors like Konrath and Eisler invested great amounts of time plugging their products, interacting with fans, exploring new markets and taking advantage of special e-tailer deals. They were shrewd marketers and masters of cross-promotions of the if-you-like-that-book-you'll-like-this-one variety. They grasped and exploited Amazon's algorithms. Above all, they recognized that publishing—self- or corporate—is a complex and full-time business requiring total focus on distribution, sales, marketing, pricing, promoting and numerous other skills.

The indie bonanza inspired countless writers to emulate the superstars, but these newcomers soon exhausted the glittering vein that the Konraths, Eislers and Hockings had mined so brilliantly. The indie marketplace was soon inundated with cheap books. As a result, consumers had trouble distinguishing good from bad, especially as some of these arrivistes produced deceptively dazzling cover art and video promos for otherwise mediocre work.

But it wasn't just readers who were confused; editors and agents too found their in-boxes stuffed with submissions that all sounded alike. Fewer authors stood out or broke away from the pack. The short happy era of self-published phenoms, boasting hundreds of thousands of passionate followers, was beginning to fade.

Agent Axelrod notes that another factor was at play. The indie culture—or more accurately, counterculture—did not easily integrate with that of traditional publishers. It was not just the dragged-out production processes and comparatively low royalties of legacy houses that self-published authors found so constraining, but also the editorial and financial supervision. The staid precincts of Big Publishers were not compatible with the rebellious spirit that characterized independent authors. And the deliberate pace of the legacy houses was completely at odds with the ability of self-published authors to move like lightning—to change the price of their e-book in minutes, for instance. Success for legacy authors was built over time; for indies it was instant and viral.

The profusion of self-published books generated growing friction, as unknown authors, unsurprisingly believing their work to be as good as that of professionals, clamored to be reviewed or at least have their books announced in industry trade publications. Unfortunately, the publishing establishment was not set up to satisfy this massive demand for attention. To publish hundreds and hundreds of reviews of self-published books, or merely to print pages and pages of notices, would add incalculable expenses to the budgets of industry publications.

But it wasn't just the cost that was so vexing, it was the rationale. Why, these editors wondered, should we pay attention to self-published books? What makes them review-worthy? Who has certified them to be of interest? What criteria are we supposed to apply to raise these books above the pack?

All these questions homogenized into a single word: "gatekeepers." Neutrally defined, gatekeepers are persons in a position to control social judgments and business decisions. Whether it be a key corporate executive or a clerk, a team manager or a clergyperson, gatekeepers are essential components of every hierarchy. In the era of social media, they sometimes take the form of "influencers," defined by SproutSocial as individuals with "specialized knowledge, authority or insight into a specific subject."[21]

For this militant band of writers, however, the definition of "gatekeeper" was anything but neutral. Suddenly the age-old assumption that literary agents and publishers were the sole—or at least the most discerning—arbiters of literary value came under question; indeed, it came under fire. As we have seen, traditional publishing was now being described by some indies as a feudalistic form of bondage; publishers were the greedy overlords, writers the suffering serfs, and agents the bloodthirsty overseers of the manor.

In a February 2014 issue of *Publishers Weekly* Mark Coker, founder of Smashwords, a major retailer of self-published e-books, wrote a blistering critique of the gatekeeper mentality that echoed the iconoclastic fervor of the Konrath-Eisler manifesto. "Authors," he said, "are losing faith in Big Publishing. Authors are angry. The moderates of the Martin Luther vein are calling for reform."

> For decades, aspiring authors were taught to bow before the altar of Big Publishing. Writers were taught that publishers alone possessed the wisdom to determine if a writer deserved passage through the pearly gates of author heaven. Writers were taught that publishers had an inalienable right to this power, and that this power was for the common good of readers. They were taught rejection made them stronger. They were taught that without a publisher's blessing, they were a failed writer.[22]

Literary agents did not escape the wrath of those hurling themselves at the gates. In an April 2009 blog post, Mary W. Walters bashed agents, calling them "talent killers" and "snakes," the vast majority of whom "do not, in fact, have any interest in literature. They are only interested in jackpots."

> Of course, an advance is no guarantee that a book will sell. But that doesn't matter to the agents. By the time the book's not selling, they already have their cuts. They simply abandon writers whose books did not hit their projected sales numbers and move on to the newest shiny thing—indifferent to the fact that they've turned those abandoned authors into the pariahs of the slush pile.

However vitriolic these declamations may be, there is more than a kernel of truth in the bloggers' righteous indignation. The dissolution of barriers, the melding of functions, the blurring of definitions precipitated by digital technology had all produced a sea change in literature, its practitioners and custodians. I myself had noted it in a 2004 blog:

> The technological breakthroughs of e-books and print on demand stunned the publishing community. It was as if the magnetic poles had shifted leaving everyone connected with books utterly disoriented as a new millennium dawned. Suddenly we were confronted by perplexing questions and paradoxes: In the coming age of disintermediation—of direct delivery of texts from author to reader—exactly what function will publishers serve? Will editors have anything to edit? Will bookstores and libraries be necessary? How will readers know what to read? Will agents be relevant? Most disturbing of all, as technology empowers authors to perform all the roles traditionally undertaken by publishers—printing, distribution, and publicity—will they still be able to define themselves as authors?

In addition to "gatekeeper," the term "discovery" was added to the lexicon of the Revolution. Formerly, readers "discovered" books by means of announcements, advertisements, reviews, publicity and word of mouth. For all but a few self-published authors, however, none of those benefits were available without serious expenditures of time, money, energy and a good measure of technical smarts. Once again, I counted myself lucky because E-Reads was principally a reissuer of previously published—and therefore already discovered—books, and we were able to promote them using the original reviews. But for anyone publishing original works in e-book format (now called "e-originals"), the challenges of discovery were difficult and onerous.

The dilemma was happily resolved by a feature on Amazon's book website: customer reviews, a democratic (or perhaps proletarian) feature that enabled anybody to be a gatekeeper.

Jeff Bezos was undoubtedly aware of the success of Yelp, the business review website founded in 2004 that utilized crowdsourced evaluations of products, services and policies with ratings on a scale of 1–5. Another successful crowdsource enterprise was the Zagat series of restaurant guidebooks. Established in 1979, founders Tim and Tina Zagat built a branded book series around the simple concept that restaurant-goers are influenced not just by professional critics but also by the recommendations of friends and neighbors. They surveyed ordinary folks asking them to rate restaurants in four categories: food, décor, service and price; the collective opinions were then summarized in annually published guides. As the company grew, it expanded its coverage to dozens of cities. In 2011, the company was sold to Google for $151 million.[23]

The crowdsource methodology perfectly suited the evaluation of self-published books, and because Amazon posted every self-published book on its website, accompanied by thumbnail cover and descriptive copy, every author got a shot at discovery and public attention. Customers were invited to post their reviews and comments, and to award the book a one- to five-star rating. Visitors could say if the review was helpful by clicking on a link beneath it.

It was a good system, but it quickly fell prey to a variety of abuses. Authors wrote their own rave reviews under false names or got friends and family to post them. A cottage industry of freelance scribes composed dazzling encomia for a few dollars each. Or, as we saw earlier, cabals ganged up to post negative reviews and award the lowest number of stars.

These practices were not confined to self-published writers. Similar transgressions were perpetrated by professional authors, too, necessitating a series of strict reforms by Amazon designed to evaluate reviewers and screen their postings. Nor were fake reviews restricted to books. Stuart A. Thompson, in a November 2023 article in *The New York Times*, asserts that the practice is a billion-dollar industry "where people and businesses pay marketers to post fake positive reviews to Google Maps, Amazon, Yelp and other platforms, and deceive millions of customers each year." Amazon, Thompson points out, "said it had blocked more than 200 million suspected fake reviews in 2022, and Google said it had removed 115 million rule-breaking reviews from Maps that year—an increase of 20 percent from the previous year."[24]

This issue continues to bedevil Amazon and similar platforms. As of mid 2025, anyone interested in creating a review or submitting "Was this helpful?" votes on Amazon needs to have spent at least $50 on amazon.com in the past twelve months and must be familiar with Amazon's community standards and guidelines. Among the prohibitions are revelation of your personal information or that of others, harassment, hate speech, profanity and obscenity, and posting advertisements. Needless to say, reviews of your own books are not allowed, nor are they permitted by friends, relatives, employers, business associates and competitors. Once postulants have met Amazon's rigorous requirements, they qualify to post comments.[25]

A phenomenally successful venture in crowdsourcing was Goodreads, the database of book information, reviews and recommendations, surveys, chats and blogs launched in 2007 by Otis Chandler and Elizabeth Khuri Chandler. Within six years of launch, Goodreads boasted 16 million members, making it an irresistible acquisition target for Amazon.[26] On March 28, 2013, Amazon announced it had bought the company.

Buyer and seller rhapsodized over the deal. Russ Grandinetti, Amazon's Vice President of Kindle Content, said, "Together we intend to build many new ways to delight readers and authors alike." Anticipating flak from readers and authors, Otis Chandler assured them that "we have no plans to change the Goodreads experience and Goodreads will continue to be the wonderful community we all cherish."

But there was plenty of flak. The Authors Guild called the acquisition by Amazon "a truly devastating act of vertical integration," according to The Guardian.[27] "With its 16 million subscribers," the Guild's president, Scott Turow, stated, "Goodreads could easily have become a competing online bookseller, or played a role in directing buyers to a site other than Amazon. Instead, Amazon has scuttled that potential and also squelched what was fast becoming the go-to venue for online reviews, attracting far more attention than Amazon for those seeking independent assessment and discussion of books."

Many users concurred with a member's comment that "unless they take an entirely 100% hands-off attitude toward Goodreads I find it hard to believe this will be in the best interest for the readers. There are simply too many ways they can interfere with the neutral Goodreads experience and/or try to profit from the strictly volunteer efforts of Goodreads users."[28]

Whether or not these fears were realized, Amazon's membership reached 140 million worldwide as of summer 2022.[29]

The website's success has not been unalloyed, however. Instances of review bombing, harassment, racist slurs and gaming of the company's algorithms have sullied its utopian goals. In December 2023, publication of a book was canceled when it was discovered that the author had created multiple Goodreads accounts in order to give other authors one-star reviews.[30] In a *New York Times* op-ed that same month, book critic Maris Kreizman wrote, "Goodreads is broken. What began in 2007 as a promising tool for readers, authors, booksellers and publishers has become an unreliable, unmanageable, nearly unnavigable morass of unreliable data and unfettered ill will."[31]

As happens in so many revolutionary enterprises over time, the free spirit of independence became more formalized, if not co-opted by establishment values, as former rebels expressed the need for structure and standards and for recognition by the public. As a result, the publishing establishment yielded to pressure to bring indie books into the mainstream.

In 2012, *Publishers Weekly* created a supplement, "PW Select," "a monthly book marketing program for self-published authors, and a supplement that appears in the magazine. Filled with feature stories about the self-publishing industry, news, reviews, book announcements listings, author profiles, and more. It was designed to provide self-published authors with a unique platform to promote their books to industry insiders, editors, booksellers, librarians, and agents."[32] PW charges authors for listing their books in its Select program.

In 2018, "Selfies" awards for self-published books were introduced in the UK and, two years later, in the United States. Among the prizes were one thousand dollars cash and five thousand worth of trade advertising; feature coverage in librarian newsletters, *Publishers Weekly* and other trade publications; and an invitation-only award ceremony. The finalists and winners were selected by a panel of judges.[33]

Because bestseller lists are compiled from bookstore sales, there wasn't and still isn't viable machinery for self-published bestseller lists unless a book is picked up by a trade house and released in traditional print. A good example is *The Martian* by Andy Weir, which was originally self-published. Crown, a division of Penguin Random House, subsequently acquired it. It then went on traditional bestseller lists and was made into a hit movie.

Ironically, the movement born out of fierce resistance to gatekeepers had created a gatekeeper system of its own. *Sic semper rebelles.*

15. THE DARK SIDE

(2000-2015)

Two thousand nine—the year that we
Were taught the benefits of "free."
A book is now considered bought
When it is sold to you for naught.
This paradox makes perfect sense
Unless you hope for recompense.
We learned that zero is a price.
If you're the buyer? Really nice!
If you're the seller? Lots of luck.
For gratis?—hard to make a buck!
It's fine for paradigms to shift
As long as authors don't get stiffed.

ALTHOUGH THE E-BOOK REVOLUTION BESTOWED countless blessings and benefits on humanity, it also unleashed a Pandora's Box of vexations. None was more devastating than piracy.

In 2010, I posted a blog that began:

> If you're lucky enough to represent Janet Evanovich, Dan Brown, Nora Roberts, James Patterson, Susanne Brockman, Lisa Kleypas, Sandra Brown or Scott Turow, you'll want to know that pirated files of your clients' books are in wide circulation all over the Internet, and they're being sold for pennies. On a typical pirate site a customer can purchase 39 Evanoviches for $8.00, 161 Nora Roberts for $13.00,

67 Pattersons for $10.00, 51 Brockmans for $9.00, 42 Lisa Kleypases for $8.00, and six Dan Browns for $4.00. The complete Harry Potter series? $7.00. Four Scott Turows? Two bucks.

Turow's case was particularly ironic because at the time, he was President of the Authors Guild. But that didn't make him less vulnerable to theft. No one was invulnerable, as is evident from that illustrious roster. "A worldwide bootleg e-book bazaar operates freely under your nose," I wrote. "Though a great many of these thieves operate out of foreign lands far beyond the reach of any law, innumerable others ply their trade domestically and defiantly, daring outraged authors and publishers to stop them—or, when stopped, reconstituting themselves like lizards regenerating lost limbs."

Although piracy is as old as art and literature, the advent of digital technology made theft and dissemination of copyrighted works a walk in the park. To savvy techies, scanning printed books, converting PDF files, cracking protected files and sharing the ill-gotten gains with an audience of millions are a cinch. Napster, the peer-to-peer music file-sharing application unleashed on the world in 1999 enabled users to download songs quickly and easily whether copyrighted or not. Before it was forced out of business and went bankrupt, it claimed some 26.4 million registered users and 6.3 *billion* minutes of use.[1] The e-book industry might have spared itself a lot of grief if it had studied Napster and hardened its defenses against piracy.

Statistics are far from reliable: Pirates do not report their gains, nor victims their losses. Nevertheless, by all reasonable accounts, the annual total losses for all intellectual property (movies, books, games, television shows, software, etc.) was in the hundreds of *billions* of dollars.[2] To give you an idea of its extent, in February 2023, a free website with *10 million* e-books was shut down.[3]

I couldn't measure E-Reads' losses from piracy, but if industry stats were any guideline, the hit on our finances was not inconsiderable. I was no longer a member of the Association of Authors' Representative; as noted earlier, I had reluctantly left the organization because of its concern about my dual and potentially conflicting roles of agent and publisher. But I used what was left of my influence to urge other agents to take action. "Downloaders," I declaimed in my blog, "whose motives range from ignorant and innocent to criminal, share files with each other, and the ripple effect of uploads can reach countless users in hours. In this Dickensian underworld, eager young thieves are rewarded with thank-you points and other bonuses, making it a badge of heroism to deliver a difficult-to-obtain trophy to their clubs."

To spread the message, I added an informational page, "Pirate Central," to the E-Reads website. There, content providers could learn about the pernicious practices of pirates and find suggestions and solutions.

Of solutions there were few. Ostensibly, the most significant weapon was the Digital Millennium Copyright Act (DMCA), signed into federal law in October 1998. It provided a framework for injured parties to file complaints and issue takedown requests to offending websites. To protect innocent third parties who inadvertently carried infringing work, however, the law created a "safe harbor" exempting them from accountability if they promptly removed or blocked access to the content upon receipt of a takedown notice.

The hitch was that under the safe harbor provision, to get stolen files taken down, you had to provide evidence that you were the true copyright owner with a valid claim of infringement. That meant digging up and copying old contracts and submitting them to the offending party, a requirement that many authors, agents and lawyers found infuriating. "The victim has to demonstrate that he or she is in truth the victim and not the perpetrator," I wrote. "Here is where injury is compounded by insult. Anyone who's ever been abused and then told that they enabled the abuse will appreciate how offensive it is for an author to be required to provide proof of authorship." Ultimately, DMCA's safe harbor ended up a refuge for the guilty.

Sometime around 2012, I learned about a company named Attributor, a piracy monitoring service founded in 2005.[4] The outfit's powerful computers crawled the Web identifying illegally posted texts on behalf of legitimate copyright owners. I likened it to giving a piece of clothing to bloodhounds searching for a missing person. After matching texts to malefactors, the company applied a variety of tactics to persuade or pressure them to take down illegally obtained content. (The firm was acquired by Digimarc in 2012.)

After employing Attributor's services to great satisfaction, I lobbied author and agent organizations to launch an allied antipiracy campaign using the software program, but these efforts met with indifference. Perhaps the Sisyphean dimensions of the challenge discouraged them. The commonly used whack-a-mole strategy for combatting pirates was pathetically inadequate, for the infestation of moles was astronomical and their appetite insatiable.

One solution that I recommended was drawn from my experience: Convert your print books and issue them as e-editions. As soon as the inexpensive e-book became available, piracy dropped off. Even if the pirated edition was cheaper than the e-book, indeed even if it was free, many consumers were afraid of clicking on sub-rosa websites that might harbor viruses or phishing traps. "Piracy springs from lack of availability," a colleague of mine, Larry Kuperman, Director of Business Development at NightDive Games, confirmed. "People will download pirated books because they're not available."

Has the problem been licked since 2010? Not in the slightest. In 2017, the Authors Guild estimated that authors lost $300 million to piracy. As the average royalty for authors is around 10 percent of publisher revenue, that would put the book industry's loss at $3 billion for that year. In 2019, England's *Guardian* conjectured that "17 percent of e-books are consumed illegally." One operator of a piracy website told *The Guardian,* "I upload anything from science fiction to ridiculously priced university textbooks. I can get any novel that I want in about 30 seconds. If I can't, I know people in my dark little corner of the Internet that can find ANYTHING that is asked for. It's incredible really." Another culprit boasted he'd pirated some 100,000 books within hours.[5]

As early as 2009, a hacker successfully cracked Amazon's DRM code, baring the retailer's files to anyone who sought them. As of this writing, you can easily find instructions on the Internet for hacking Amazon, and if they're not clear you can watch a YouTube video to make sure you perform the task correctly.[6] Just be careful. Pirate downloads lead to 36 percent of phishing theft and malware infections, costing up to $114 billion annually.[7]

It isn't just the ease of stealing that has saturated the book world but the ease of conscience among downloaders—or should I say freeloaders—that justifies obtaining books without paying for them. *The Guardian*'s survey of readers who did so yielded many creative, if not tortuous, rationales for what authors would plainly call theft. Many respondents pleaded poverty, explaining they loved books but couldn't afford them. Others differentiated the shoplifting of clothes or groceries (bad) from the boosting of book files (okay).

Some reasoned that the authors of bestsellers were so rich they wouldn't miss a purloined copy or two, an argument echoing the vituperation hurled at the heavy metal band Metallica after its music was flagrantly stolen by Napster in 2000. The band not only sued Napster, it also demanded that the company block over 350,000 users, a move that provoked the outrage of fans. Victim though the band clearly was, fans called its members greedy millionaires. "Metallica were no longer countercultural icons," kerrang.com reported, "but fat-cat faces of an outdated music industry."[8]

Still other downloaders claimed, in *The Guardian*'s words, "a greater ethos of equality, that 'culture should be free to all.' "

This last excuse was not just a clever way of salving a guilty conscience. It was a precept espoused by a libertarian element that believed the protection of intellectual property was unsupportable and must be eradicated. The phrase "Information Wants to Be Free," attributed to Stewart Brand, founder of the *Whole Earth Catalog,* was uttered at a hackers conference in 1984, according to R. Polk Wagner, Assistant Professor at University of Pennsylvania Law School, writing in the *Columbia Law Review.*[9]

Though the expression was ambiguous, it was seized by some intellectuals as a means of justifying the "liberation" of protected information, literary works and computer source code from the control of patent and copyright owners, who were seen as unbridled capitalists and rapacious exploiters. The philosophy (if it can be called that) appealed to those who had grown up in a culture of free content and felt entitled to download and share files that were there for the taking. One articulate advocate, Cory Doctorow, posted a defense of the movement in a May 2010 issue of *The Guardian*. Here are some of his arguments in support of IWTBF advocates:

- They want open access to the data and media produced at public expense, because this makes better science, better knowledge, and better culture—and because they already paid for it with their tax and license fees.
- They want to be able to build on earlier creative works in order to create new, original works because this is the basis of all creativity, and every work they wish to make fragmentary or inspirational use of was, in turn, compiled from the works that went before it.
- They want to be able to use the network and their computers without mandatory surveillance and spyware installed under the rubric of "stopping piracy" because censorship and surveillance are themselves corrosive to free thought, intellectual curiosity and an open and fair society.

Call it emancipation or call it larceny, that sense of entitlement has contributed to the estimated millions lost annually by book publishers due to e-book theft and a 24 percent drop in author income from 2013 to 2020.[10] The malady has afflicted every field of endeavor. Techjury.net, a team of software experts that collects statistics on piracy, asserts that 34 percent of Gen Z use stream-ripping, resulting in a projected $67 billion in lost video revenue in 2023 worldwide. *The Recording Industry Association of America* estimates music theft leads to the loss of $2.7 billion in earnings annually in both the sound recording industry and in downstream retail industries, with a concomitant loss of $422 million in tax revenues.[11]

Plagiarism is a cousin to piracy, and in terms of the pain inflicted, its victims will tell you there is no difference between the two. In a blog on writing-world.com, Jack Lynch tracks the origin of the word to the Roman Empire:

> When the Roman poet Martial accused a rival, Fidentinus, of stealing his verses, he called him a "kidnapper"—in Latin, *plagiarius*. The term stuck. The Latin word made its way into English in 1601 when Ben Jonson described a literary thief as a *plagiary*, a word Jonson's near-namesake, Samuel Johnson, defined in his *Dictionary* of 1755 as "A thief in literature; one who steals the thoughts or writings of another" and The crime of literary theft."[12]

The advent of AI has thrown the issue into a new and uncharted dimension. According to a 2021 report in THE [Technological Horizons in Education] Journal by Dan Schaffhauser, "Plagiarism among students jumped by 10 percentage points after the 2020 Covid-19 pandemic, when classes went online—an increase in the average rate of copying in student work from 35 percent to 45 percent."[13] Transgressors who used obscure source material in the expectation of escaping scrutiny can no longer hide, for artificial intelligence can instantly locate purloined sources no matter how remote. Instances of leading figures accused of plagiarism abound in academia, law, literature and politics. In the publishing field, the offense is becoming a plague. Editors have been inundated with suspect submissions requiring special tools to discover research sources. Some publications require authors to affirm that they have not used AI to research or write the text they are submitting. There are AI tools for detecting AI plagiarism.

The underlying dynamics of AI have actually made it harder to define plagiarism. On behalf of its students the University of South Florida asked ChatGPT to define the term. It answered:

> Plagiarism is defined as the act of using someone else's work or ideas without giving proper credit or attribution. In the case of using

answers generated by chat AI technology, it can be argued that it is not plagiarism because the answers are not the work of a single individual, but rather the result of a complex algorithm that draws from a vast amount of data and sources.

The bot went on to say (in bold print for the benefit of the school's students):

> However, it is important to note that using chat AI technology to generate answers without proper citation or attribution could still be considered unethical or academically dishonest, particularly if the user is trying to pass off the answers as their own original work. In academic settings it is generally expected that students and researchers provide proper citation and attribution for any sources they use, whether those sources are generated by chat AI technology or not.[14]

As I type these words, a lawsuit is in process, brought by several authors and the Authors Guild, alleging that OpenAI, a company using generative artificial intelligence to train its chatbots, is infringing on the authors' copyrights. In essence, the complaint states that the company utilized the plaintiffs' texts without their permission, and that the chatbots are capable of producing derivative works that simulate the authors' books, depriving them of potential income. In addition, *Publishers Lunch* reports that "over 10,000 [actually, 11,500] creators and organizational partners from around the world have signed on to a simple statement of protest, declaring: 'The unlicensed use of creative works for training generative AI is a major, unjust threat to the livelihoods of the people behind those works, and must not be permitted.' " The release went on to say, "These consumer-facing [AI] models and tools would not exist without the books, newspapers, songs, performances, and other invaluable human expressions that were—and continue to be—copied, ingested, and regenerated in blatant disregard of the law."[15]

Though the issues raised by AI appear to be unprecedented, in fact all of them were litigated over a decade ago. Sadly, in *The Authors Guild Inc., et al. v. Google, Inc.*, it did not go well for authors.

In 2004, Google announced an enterprise called the Google Library Project. The company had approached five universities about scanning their collections to create a searchable database that included both copyrighted and public domain titles. Though Google (whose motto was "Don't be evil") copied the complete books, it displayed only "snippets" on the Web.

A year later, the Association of American Publishers as well as the Authors Guild filed separate lawsuits against Google for copyright infringement. The suits were presently consolidated.

In its defense, Google asserted that its actions were sanctioned by the fair use provision of the United States Copyright Act of 1976, as amended in 1992. That provision protects unauthorized users if the use can be demonstrated to be for the purposes of education and research. Although the law employs a number of criteria to analyze a work's use, the key factor was whether Google's application of the texts was "transformative." That is, did it alter the function of the work in a way that the author did not contemplate?

As Adam W. Sikich, Senior Counsel at Dunner Law PLLC, summed it up, "Google would argue that providing an online searchable database of millions of books (including many that are out of print) available to the world is highly transformative because such a use promotes research and knowledge." Sikich added, "Google may even claim that by providing online access to information about the digitized books, the Google Library Project may actually enhance the marketability of copyrighted books."[16] Sergey Brin, cofounder of Google, went so far as to declare that his company was rescuing the world's cultural treasures from doom. "The famous library at Alexandria," Brin warned, "burned three times . . . as did the Library of Congress. . . . I hope such destruction never happens again, but history would suggest otherwise."[17]

A settlement was reached that would enable Google to continue operating its program if it compensated authors whose copyrighted works it had scanned. The agreement also allowed authors and publishers to opt out of the Library Project. Unfortunately, the compromise was challenged over a host of perceived inequities, and after years of futile attempts to create a new settlement that would make everybody happy, negotiations collapsed and the parties went to court.

In November 2013, in a summary judgment, Judge Denny Chin ruled for Google, proclaiming:

> In my view, Google Books provides significant public benefits. It advances the progress of the arts and sciences, while maintaining respectful consideration for the rights of authors and other creative individuals, and without adversely impacting the rights of copyright

holders. It has become an invaluable research tool that permits students, teachers, librarians, and others to more efficiently identify and locate books. It has given scholars the ability, for the first time, to conduct full-text searches of tens of millions of books. It preserves books, in particular out-of-print and old books that have been forgotten in the bowels of libraries, and it gives them new life. It facilitates access to books for print-disabled and remote or underserved populations. It generates new audiences and creates new sources of income for authors and publishers. Indeed, all society benefits.[18]

The ruling was appealed all the way up to the Supreme Court, which declined to hear it, leaving Chin's decision to be the final word on the matter.[19] Wholesale scanning of complete copyrighted works is the law of the land, for—presumably—"all society benefits."

Tell it to the authors.

16. INDELIBLE INK

(2000–2015)

Publishers expressed enchantment
With the notion of enhancement.
Audio, video, music, flix,
Bangles, baubles, Bar Mitzvah pix.
A tune or two was all it took
To constitute a mobile vook.
They tossed in every kind of crap
And designated it an app.

NOTWITHSTANDING PREDICTIONS BY E-BOOK PIONEERS of a paperless world; notwithstanding denunciations by indie anarchists that the feudalist book business was on the verge of collapse; notwithstanding publishers' struggles to adapt their superannuated industry to a dazzling new technology—the print book industry did not founder. In fact, it went on to thrive and thrives to this day. According to Statista, "Print . . . remains the most popular book format among U.S. consumers, with 65 percent of adults having read a print book in the last twelve months."[1] While e-books maintain a significant place on the reading spectrum, they have proportionately lost ground since their debut early in the 2000s.

What accounts for print's unshakeable traction? Or to reverse the question, why haven't e-books conquered the paper world as the technology's progenitors predicted? There are several answers.

My own theory is that a whole generation of kids grew up reading printed books or having them read to them or selecting them on visits to the library. Thus, their preference for that format carried over to adult reading. In the dawn of the digital revolution, publishers experimented with hypertext and hotlinks to produce animated books that children could not only read but watch and hear as well. These books "read" themselves without parental intervention. The idea was for children to learn to read on their own through stimulating images and sounds, vocabulary prompts that pronounced the words aloud, pictures that moved and danced and flashed. These gadgets, subsequently called "vooks" (a portmanteau blend of "video" or "virtual" and "books"), would liberate parents from the task of helping their children to read and learn. In 2009, Vook, a software platform for the creation of digital video books, was introduced.[2]

This was great in theory, but it failed to recognize that few parents believe reading to their children is an unpleasant task. Quite the contrary, they love snuggling up with their kids and reading together. The intimacy of reading a print or picture book to your child, or having your child read to you, cannot be duplicated by self-reading devices. As I wrote in 2012, "Though picture book apps and stories that 'tell' themselves without parents present are great fun, they just don't seem to have the same appeal as the warm body and familiar voice of mommy or daddy."

It is arguable (based on two thousand years of human interaction with the codex format) that the tactile experience of holding and reading a printed book is superior to the awkwardness (for children at least) of manipulating text and pictures on a computer.

There was something else: Many parents sensed that children do not benefit from reading on screen as much as they do from immersion in print books. A number of studies confirmed that children are easily distracted by e-books and do not retain information the way they do from printed ones. One such experiment, conducted by a team headed by Tiffany G. Munzer, MD, Department of Pediatrics, University of Michigan Medical School, confirmed what previous studies had found: "less dialogic interaction between parents and preschoolers during electronic-book reading versus print." They concluded:

> Parents and toddlers verbalized less with electronic books, and collaboration was lower. Future studies should examine specific aspects of tablet-book design that support parent–child interaction. Pedia-

tricians may wish to continue promoting shared reading of print books, particularly for toddlers and younger children.[3]

As children matured and became adept with computers and cell phones, the e-book became an option rather than a necessity, good for some kinds of reading but not the device of choice for settling in with an immersive story or researching a paper.

Children's preference for *book* books over electronic ones has carried into the present generation and been passed on to the next one, as evidenced by the perpetually strong market for children's books. Though sales of adult books have fluctuated in the twenty-first century, print books for young readers have held the line and made a profit year after year. During the boom years for e-books of 2008–12, sales of print trade books fell 8.4 percent except for one category: children's books. *Publishers Weekly* noted that in that period "children's/young adult had the strongest gain, with sales jumping 117 percent, from $215.9 million to $469.2 million."[4] Sales of children's books have continued to soar, reaching some $3 billion in 2022.[5]

Another successful boost for printed books comes from the area of expensive special editions, artistically designed formats that fans and collectors can display on their bookshelves. Limited editions have always been popular in the science fiction genre but as the term "limited" implies, the printings were deliberately small to keep the value of these rare volumes high. However, in the early 2020s, special editions became all the rage in romance as well, with one publisher issuing printings exceeding a million copies. As Alexandra Alter described them in *The New York Times*, "Publishers are investing in colorful patterned edges, metallic foil covers, reversible jackets, elaborate artwork on the endpapers, ribbon bookmarks and bonus content."[6]

Print's sustained hardiness was also reflected in the paperback sector. On the bad-news side, sales of the mass market format have been steadily declining over the past two decades. Their death notice was recently announced in *Publishers Weekly*: "Sales of mass market paperbacks have steadily declined in recent years, to the point where they accounted for only about 3 percent of units sold at retailers that report to Circana BookScan in 2024. The format will take another big blow at the end of 2025, when Readerlink will stop distributing mass market paperbacks to its accounts."[7] ReaderLink describes itself as "the largest full-service distributor in North America" with six U.S. distribution centers supplying over 100,000 stores.

The good news is that the slack has been taken up by the success of trade paperbacks.[8] Major publishers are shifting their focus to trade paperback as the format of choice both for originals and reprints. Even mass-market paperback publishers that prospered with genre literature like romance and science fiction are pushing their chips onto the larger trim size.

This seismic shift is not just a matter of taste but also reflects the drastic change in the way books are distributed and displayed. The old mass market system—monthly selections delivered to candy store, grocery, and drugstore racks—has been largely replaced by bookstores whose shelves are better designed to stock and display trade paperbacks. This format is more economical in reprinting hardcovers because they are the same trim size and often use the same cover. The average return rate for trade paperbacks is considerably lower than the 40 percent or more for mass market paperbacks, because trade paperbacks have far longer shelf lives.

Trade paper has become the preferred format for literary originals and reprints of most trade hardcovers, whereas mass market paperbacks are for the most part reserved for reprints of major bestsellers. (The exceptions are books in genres like science fiction, romance, westerns and horror, which continue to sell as mass market originals and are carried in big-box outlets like Walmart.)[9]

Another reason for the popularity of trade paperbacks is that when they go out of stock, publishers can replenish them using print on demand, which reduces the risks both of overprinting or underprinting. In fact, print on demand is another reason why print books have prevailed. Books that used to die after their initial printings have a long, if not infinite, life thanks to POD technology. Although the term "long tail" was invented to describe a somewhat different business strategy, I felt that it applied perfectly to the infinite life of digitized books—and to the power of POD to make niche products like specialized books available to almost anyone, anywhere.[10]

As for my prediction that the returns-driven publishing business would be supplanted by one based on print on demand, I got it half right. By 2020, LightningSource, the leading on-demand printing company, had grown into a $2 billion company, with 18 million titles in its inventory.[11] But the consignment model, that relic of an era when returns were modest, endures, especially when it comes to blockbuster books, where economies of scale more than balance losses from returns. Publishers are better able to manage printings, distribution and inventory than they were in the Roaring 1990s. They simply regard losses from returned copies as an inevitable cost of doing business and pass them along to consumers in the form of higher list prices.

Another boost to the fortunes of printed books was assimilation of the new media. By the second decade of the twenty-first century, a young, technically astute generation of editors had figured out how to apply to book production solutions developed by e-book publishers and self-published authors. The result was improved marketing, targeted advertising and publicity, robust social media campaigns, and greater synergy between print and other media, such as audiobooks. In his 2023 end-of-year message, Hachette CEO Michael Pietsch noted that "more than half our sales take place on digital retail platforms. . . . With new tools that track how often customers look at a book page and then go on to purchase the book, this program encourages our publishing teams to measure and adjust selling lines, title descriptions, images and other elements of our book pages in order to drive more sales."[12]

These improvements may explain the paradox that the number of book publishing jobs has dramatically declined in the past 25 years, from 91,100 in 1997 to 54,822 in 2023, according to the U.S. Bureau of Labor Statistics.[13] (This statistic doesn't include people employed in self-publishing, which is impossible to quantify.) A 2024 *Publishers Weekly* article by Thad McIlroy and Jim Milliot confirms that "publishing has become more efficient in the digital age." Yet profitability has improved, confirming the sustained strength of the traditional book business—what book developer and agent Philip Turner described as "The Persistence of Print."[14]

Publishers created departments dedicated to producing in-house media that they had formerly farmed out. Formats such as e-book, podcast and audiobook that used to be optional in book contracts became deal-breaking must-haves.

A host of conferences, conventions and book fairs helped to translate complex technical ideas into practical publishing operations. Two in particular provided opportunities for the traditional and technological communities to exchange DNA and interact in what Tim O'Reilly, founder and CEO of O'Reilly Media, called "the reinvention of the publishing industry."

The Tools of Change conference was launched in 2007 by O'Reilly; Digital Book World was begun in 2010 under the dual leadership of Michael Cader, founder of the *Publishers Lunch* industry newsletter, and the late (and greatly lamented) Mike Shatzkin, author and book industry strategic consultant. These convocations were well attended both by old-guard publishing people and eager young editors. Their programming included speeches, panels, workshops and demonstrations of new technological tools, plus opportunities to network and brainstorm during session breaks, lunches, cocktail hours and parties.

Though Tools of Change was discontinued in 2013, Digital Book World continues to this day. Its 2023 agenda covered such futuristic topics as artificial intelligence and nonfungible tokens.

As we have seen, book houses became more aggressive about going after successful self-published authors, a number of whom were happy to trade the daily grind of managing their titles for a big advance, the prestige of association with a name-brand publisher and the emoluments of trade book publication, such as advertising and serious review attention. As one author confessed to me after jumping from self-publication to traditional, "Now I appreciate why publishers get ninety percent and authors ten."

Another gambit was for trade book publishers to acquire successful self-publication companies. The biggest of these was Author Solutions. Founded in 2007, the company boasted 190,000 titles written by 150,000 authors and eventually scooped up several competitors including AuthorHouse, iUniverse and Xlibris.[15] In 2012, Pearson (then owner of Penguin Books) acquired Author Solutions from Bertram Capital Management for $116 million.[16] When Penguin merged with Random House in 2013, Author Solutions became a division of that colossus.

Author Solutions also figured in another significant deal with a major publisher. Late in 2012, Simon & Schuster announced a partnership with that company, forming Archway Publishing, dedicated to assisting writers. "S&S will refer authors who submit unsolicited manuscripts to the Archway program and will monitor the success of Archway titles," *Publishers Weekly* reported. "S&S hopes to differentiate Archway from other self-publishing programs by offering some exclusive options that include a 'concierge service' in which authors will have the chance to work with a dedicated publishing guide who will coordinate each step of the book production process." The enterprise promised to be profitable, with fees ranging from $1,999.00 to as high as $24,999.00.[17] (In 2016 Penguin Random sold Author Solutions to a private investment firm.)[18]

If print books prospered in the twenty-first century, Book Expo America, the American book industry's biggest annual event, should have prospered, too. Paradoxically, the conference went into decline and died in 2020. Why? Possibly because BEA was essentially an analog island in a digital ocean, fatally slow to embrace the new technology even during the e-book's most expansive period from 2009 to 2020.

The American Booksellers Association convention had started in 1947 but was renamed Book Expo America when Reed Exhibitions acquired it in 1995. Independent booksellers came to New York City from all over the country and indeed all over the world, to Manhattan's mammoth Jacob Javits Center, to "crawl" the aisles and gawk, meet authors, gossip with old friends, check out trends, collect catalogs, grab ARCs (advance reading copies) of big books and order titles for the coming season from exhibiting publishers.

BEA was always about the celebration of trade books, their publishers and authors, and few digital companies could be found in the roster of exhibitors. *Publishers Weekly*'s coverage of the 2012 Expo put it this way:

> This year at BEA publishers are once again handing out galleys, bring-
> ing hopeful debut authors to sign and schmooze, and parading their
> bestsellers, believing that lightning will strike twice, or three, or 50
> times. We're crazy about all of it because despite the wonders of dig-
> ital and the brave new world, there's nothing like the excitement of
> watching an author finding the page to sign, or the fans waiting to
> meet that author whose book has broken their heart.[19]

Yet, for all its bookish glamour, BEA just wasn't sexy enough to maintain its attraction, and attendance began to drop.

Sensing this deterioration, the show's sponsors tried to shore it up by opening it up to fans in 2014, a move aimed at infusing a younger and more energetic vibe into the staid proceedings, especially in the area of young adult books, whose authors were greeted with shrieks of adoration and swarms of fans lining up to meet their idols and get their autographs on advance reading copies. That lasted a few years, but eventually, enthusiasm for BookCon, as the new feature was called, waned.

BEA lacked the go-go excitement of digital publishing, which had become an organic component of the book industry. The cost and hassle of travel to New York City was also a factor. Booksellers were relying more and more on remote means of studying publishers' online lists (called "e-catalogues"), reading PDFs of forthcoming books, communicating with their sales reps and ordering books. Attendance had drastically declined from 13,872 in 2010 (not counting exhibi-tor personnel) to a mere 8,260 in 2019.[20] Sadly, the Covid pandemic finished BEA off, as it did so many other in-person events.

Online programs were substituted, but because the Expo had been struggling with attendance and identity issues anyway (and the number of days it was open was down to two), it was decided in 2020 to "retire" BEA until a new format and approach could be found to lure the book community back into convention centers.[21]

BEA was about traveling to a geographical locus to celebrate physical objects, the very definition of an analog event. It died of irrelevance. R.I.P. BEA.

17. FOR SALE

(2013)

In the beginning was the word,
Once was read but now is heard.
The audiobook now reigns supreme
On CD, tape cassette and stream,
Headphones, ear buds, iPods, jacks
Have all replaced mass paperbacks.

I N 2013, *PUBLISHERS WEEKLY* DECLARED that "the 2008–2012 period certainly qualifies as the boom years for e-books, a period during which the format moved from something of a curiosity to a vital part of the publishing industry."[1] As I hope this book has demonstrated, e-books were scarcely a mere curiosity! But the term "boom" for that period is completely apt, as E-Reads' ledgers confirm. Indeed, 2012 was our biggest year ever.

Yet I began to detect signs of softness in the marketplace, and my bones told me it might be time to consider getting out of the e-book business. It wasn't long before I had hard confirmation. Industrywide sales had leapt in triple-digit percentages between 2008 and 2011, but the growth from 2011 to 2012 was only 44 percent. Of course, "only" is a bit deceptive, as sales were now in the billions ($2.1 billion in 2011), and it was unrealistic to expect triple-digit billion-dollar jumps year after year. But that was just one of several symptoms of a slowdown. Sales of Barnes & Noble's Nook descended from a pinnacle of $933.47 million in 2012 to $780.42 million the next year, the start of a plunge to $92.14 million in 2019.[2]

Sales of another E-Reads stalwart, Sony e-book, were dropping annually. (They finally sold their e-book division to Kobo in 2014.) The real fire alarm sounded when our Kindle receipts fell by 30 percent between 2012 and 2013, from $827,424 to $575,905. As a result of these setbacks, we took a 20 percent hit in revenue in 2013, $1,200,000 compared to the $1,500,000 we had grossed at our 2012 zenith. A million-two was still a lot of money (for us, at least) but if it meant we were starting to slide down a slippery slope, I had to consider winding the operation down or selling it altogether.

Several other factors bolstered my pessimism, especially piracy. Despite all of the book industry's heroic measures, the pirates were eating us alive. Many of our precious copyrighted works were easily obtainable on illicit websites if you knew where to look and weren't afraid of getting scammed or infected. Many consumers had downloaded enough ill-gotten books to satisfy five lifetimes of reading and shared them with God knows how many people, so that ruled out all those potential customers.

Another factor was audiobooks. After Amazon's acquisition of Audible in 2008, this formerly niche medium had begun its dramatic ascendancy. Though reading and listening are two completely different mental functions, the criss-crossing of audiobooks' rise and e-books' descent was obviously having an impact on reading culture. The advantages were obvious: You couldn't read a book or e-book walking to work or while driving, but you could listen to an audiobook. I myself had gotten hooked on the pleasures of listening after I got bored on a treadmill watching highlights of last night's football game. I discovered a wonderful website, librivox.org, and consumed tons of classics I had neglected in my dissipated youth.*

If others enjoyed them as much as I, audiobooks were going to give e-books a run for the money.

As we now know, they did.

One more factor weighed heavily on my mind: the competition. As I mentioned, it had become easy to start an e-book/POD publisher and offer a bigger royalty than the 50 percent we were paying. Though I didn't feel threatened by most of these start-ups (or should I say upstarts) and our authors were loyal and

---

* The downloads were free because the books were all in the public domain and the readers all volunteers. As a result, Librivox incurred neither licensing nor talent costs. The website is supported by donations.

seemed happy, there was no guarantee that some of them wouldn't be tempted to jump ship when their contracts with us expired.

I couldn't divine all the underlying social and cultural factors causing the decline in e-book sales, but I knew down from up, and down was where the e-book business seemed to be pointed in 2013.

Of our many rivals, the most formidable was Open Road. Though the company was launched ten years after E-Reads, Open Road's coffers were loaded with investor bucks and they were on an acquisition spree. CEO Jane Friedman's VIP connections generated a glamour and electricity that our unassuming little operation could not hope to match. Plus, her company promised to promote its authors, a benefit that was out of the question on our frugal budget. Late to the party though they were, they'd made a big splash and were attracting a lot of significant authors and estates.

Though I felt threatened by Open Road, I recognized that they could be instrumental in breaking open a market populated by the hesitant authors and skeptical agents who had sat out the first wave of the Revolution, the ones I called VLAs—Very Late Adopters. The good news (for Open Road, at any rate) was that by the time Jane's company opened its doors, the e-book industry had begun to stabilize. A recognizable business model was more or less in place. It was becoming clear to these distinguished holdouts that e-books could breathe new life into their fading works and new revenue into their fading bank accounts.

Jane and I were friends and had done business together. Particularly memorable was our deal on a book about beer by Garrett Oliver, brewmaster of Brooklyn Brewery. Instead of a conventional pitch session, I arranged for Jane to host a beer tasting in the conference room of HarperCollins. Some two dozen executives sampled cheeses and chocolates with a variety of beers about whose pedigrees and flavors Garrett rhapsodized from a podium. A really, *really* good time was had by all with many refills and raucous toasts. The next morning Jane bought the book.

She and I had stayed in touch since the founding of Open Road, and as time wore on, we conducted a kind of professional flirtation suggesting we might eventually end up working together. I knew she was particularly keen for our science fiction list, which in one stroke would make her company a major player in that genre.

Early in 2013, she invited me to her elegant club, where she expressed admiration for E-Reads' books and I expressed admiration for Open Road's wealth.

She sounded me out about acquisition of our company, but my calendar shows a six-month hiatus, indicating that I took my time thinking about it. In July of that year, however, my log shows another invitation to her club, and this time I was prepared to succumb to her blandishments if the price was right.

Jane asked, "How do you want to do this?"

I said, "You name a number, and I'll laugh. Then I'll name a number, and you laugh. Then we'll get serious and work out a deal." She did and I did and we did. It included engaging me as consultant for a period of time to help with the transition. A clink of glasses, kisses on both cheeks and the deal was done.

Months of due diligence followed, the mind-numbing process in which the acquirer scrutinizes the acquiree's contracts and financial records and ascertains their legal *bona fides*. I would rather count a vault full of nickels than do due diligence, but we had maintained impeccable records, and there were no serious hitches.

Anxious that our staff would not abandon me before it was absolutely clear that the deal would go through, I performed the diligence secretively. Sometime late in 2013, Open Road's legal counsel declared that everything was in order. It was time to tell my crew.

They were not completely surprised, as they too had sensed a weakening of the market or perhaps of my enthusiasm. They pledged to stay on to the end. I loved these people. We had forged a strong and proud team. Fearlessly, they had applied their unique skills to blaze trails through the dark and tangled terrain of a medium that was still inventing itself. Unlike many shops that batted e-books out as fast as they could produce them, we had approached each book as a challenge that required personal attention and hand-crafting. The finest compliment paid to us was from an author who described E-Reads as "the artisanal bakery of the e-book industry." Our diligent attention to detail, relentless pursuit of editorial perfection, scrupulous accounting and lucid communications with authors were held up as a model. Each e-book and POD we turned out felt unique. And here I would like to give a special shoutout to the late John Douglas, our meticulous contracts administrator, a witty and charming man who was taken away from us far too early.

Naturally, I had to tell the authors. Some of them were clients of other agencies, but most of them my own clients. Though our agency didn't take commissions on E-Reads royalties, I had remained involved in their lives and careers, selling their frontlist books to traditional publishers and issuing their out-of-print books in E-Reads. The news of the sale came as a jolt to many of them, and they had a million questions. I reassured them that Open Road would make them even more money than E-Reads was able to do, and besides, I had been retained by the company for several years and would always be available to address their concerns and issues.

The launch of our titles under the Open Road banner was scheduled for April 1, 2014, requiring us to dispel speculation that it was an April Fools' Day prank. We uploaded all our files to their computers, and after a host of tests, we hit Enter, ending E-Reads' fifteen year existence.

In retrospect my instinct to sell the company was sound, for the e-book business did indeed soften after E-Reads' peak years. My confidence that Open Road would provide a happy home for our authors turned out to be well placed. In the following years, their royalties rose thanks to Jane Friedman's "secret sauce," a proprietary program designed to optimize the marketing of its titles. So successful was this platform that Open Road attracted some seventy other publishers to use it to promote their own books. Between 2018 and 2021 Open Road's sales doubled. In 2021, the company was sold to a group of investors for somewhere between $60 and $80 million.[3]

18. RESCUED

(2000–2013)

UNDERLYING E-READS' BUSINESS PLAN was the simple conviction that there was a commercial market for the books we chose to reissue. Those works were once considered by their original publishers to be good enough to see the light of day, but in time, sales dwindled or ceased altogether, and their publishers decided not to reprint them. Now I had to make the same decision: Should I give them a second life? Was there a new readership out there to profitably support reissue?

The answer was a resounding *yes*, and it was gratifying to see how passionate readers were to discover or rediscover books that had been drifting into oblivion.

Here are some authors whose luster was restored.

TRISTAN JONES

It happens that when I was younger, I developed a fervent but totally incomprehensible love for books about the sea. I say incomprehensible, because my only relationship to oceans has been to bathe in them. Yet the lure of the sea, however vicariously satisfied in books, has held an immense power for me. I've read countless novels and true accounts of men locked in mortal combat with this most formidable of elements. It was this passion that led me to read all of Tristan Jones's books. And it was the experience of immersing myself in his astonishing sailing adventures, narrated in his ravishing and lilting prose, that impelled me to write him a fan letter.

Not long afterwards, my secretary announced that Tristan Jones was on the phone. I wondered from what exotic locale this ship-to-shore communication came. The caller identified his position as a payphone at latitude Seventh Avenue, longitude West 8th Street, in Greenwich Village, on the sun-drenched isle of Manhattan. My hero was a fifteen-minute taxi ride from my office! By the end of that fifteenth minute, the crusty old salt was sitting across my desk from me. We went out for a drink—indeed, many, many drinks. Three hours later he bestowed his e-book rights upon E-Reads.

High on the list is Tristan's great original trilogy of seagoing adventures, *Saga of a Wayward Sailor*, *Ice!* and *The Incredible Voyage*. Though his stories sometimes push the limits of credulity, they are forgivably delicious in the telling.

COLIN COTTERILL

One day in 2003, I received a letter from Thailand containing one of the oddest pitches I'd ever read. The uneven and ink-blotched fonts signaled it had been composed on a manual typewriter of somewhat hoary vintage. The query described a mystery novel, featuring a septuagenarian coroner in Communist Laos in 1976, populated by a bizarre cast of characters and set in a spooky, obscure milieu. Could anything be further from commercial? What stayed my hand from tossing it was the wit and charm with which the author, Colin Cotterill, confessed in his letter that he *knew* how unusual his description must sound and how inadequate he was at synopsizing the story. He promised the book was far better than his summary.

Okay, he hooked me. I sent for *The Coroner's Lunch*, and in time, the manuscript arrived.

It blew me away.

The pitch was as hard for me to write as it had been for Colin. Not surprisingly, the major publishers didn't get it. However, browsing through *Publishers Weekly*, I came across an article describing Soho Press's predilection for mystery stories set in exotic locales. I reached out to the editor, writing, "You want exotic? It just doesn't get more exotic than this book!"

She, too, loved it and bought it, the first of a series currently numbering fifteen. Fortunately for us, Soho did not understand e-books and permitted E-

Reads to retain the rights *The Coroner's Lunch* became a dazzling international success. It was a *Booklist* Book of the Year. *Kirkus Reviews* gave it a starred review saying, "This series kickoff is an embarrassment of riches: Holmesian sleuthing, political satire, and [a] droll comic study of a prickly late bloomer." *The Orlando Sentinel* said, "In Siri, Cotterill has created a detective as distinctive as Maigret or Poirot."

And E-Reads had the e-book rights to it and the first sequel, *Thirty-Three Teeth*. Eventually, we relinquished them to Soho, but not before collecting a lot of money from that unfamiliar medium called e-books.

Of course, launch your reading with *The Coroner's Lunch*.

EMILY HAHN

Should I end up going to Heaven and be granted admission to the Literary Division, I will make a beeline for Emily Hahn. Though *The New Yorker* described her as a "forgotten American treasure," E-Reads helped to keep her name alive by issuing nine or ten of her fifty-four books, works that spanned a universe of subject matter so diverse that her publishers didn't know how to categorize her. I'm not sure anyone could, with a bibliography ranging from *Aboab: The First Rabbi of the Americas* to *Seductio ad Absurdum: The Principles and Practices of Seduction—A Beginner's Handbook*; from *Leonardo da Vinci* to *Mary, Queen of Scots* to *Chiang Kai-Shek*; from China to India to Singapore to England to Africa; from adult to children's; from historical to contemporary.

Mickey—as she was known to her intimate crowd—was born a fully formed and extraordinary person. She might be described as a proto-feminist, for it doesn't seem to have occurred to her that being female was a disadvantage that needed to be overcome. She blazed trails hitherto untrodden by many other women. She earned a degree in mining engineering in an all-male class; she traveled abroad unchaperoned in male bastions; like George Sand before her, she wore trousers and smoked cigars; she explored the Congo unaccompanied; she conducted an intimate relationship with a Chinese poet; she bore an out-of-wedlock child with a married British officer (whom she subsequently married); trapped in China when the Japanese invaded it in 1941, she negotiated her freedom in exchange for English lessons given to the country's occupiers.

The title of Ken Cuthbertson's biography of her sums it all up: *Nobody Said Not to Go*. Nobody dared tell Emily Hahn, "You can't do that." There was nothing she could not do. There was little she didn't try to do. There was scarcely anything she did *not* do.

Emily's books recounting her stirring Chinese sojourn, including *China to Me* and *The Soong Sisters*, have been acquired by a motion picture company. Could one ask for better testimony to the recovery of out-of-print literature?

DAN SIMMONS

Harlan Ellison had a bloodhound's nose for detecting great writers, and he referred me to such masters as George Alec Effinger, Poppy Z. Brite and Dan Simmons. Harlan would phone me any hour of day or night declaring, "You've *got* to represent this fantastic writer! Here's the number! Call now! I told them you're great! Call me after you talk to them! Don't screw this up!"

He brought Simmons to my attention after the thirty-four-year-old writer won first prize in a short story competition, but even Harlan's hyperbole inadequately described Simmons's brilliance. His first novel, *Song of Kali*, won the World Fantasy Award. *Hyperion*, the first saga in what was to become a classic science fiction quartet, won the Hugo, science fiction's highest honor. His heart-stopping horror thriller *The Terror* won the Horror Writers Association's Bram Stoker Award and was made into a ten-part television miniseries. Over the next twenty-five years, he was to collect innumerable awards in fantasy, science fiction and horror, including eleven conferred by *Locus*, the science fiction trade magazine. And by the way, he also wrote a masterful trilogy of thrillers featuring a hard-as-diamonds protagonist, Joe Kurtz.

Dan's versatility gave his publishers fits. For the sake of branding, they urged him to keep to one genre, but Dan followed his bliss and wrote whatever seized his imagination. I once suggested to him he might do better by sticking consistently to one category or another, and he let me know with both barrels that his job was to write 'em and mine to sell 'em. I never made that mistake again.

We became tight friends, a rare blend of personal and professional. Whenever he came to New York, we explored stories, characters and publishing gossip long into the night over a bottle of fine twenty-five-year-old malt whiskey, and snatches of our colloquies showed up in his books. So strong was our bond that he honored me by dedicating a book to me.

Dan, Harlan and I made innumerable plans to get together; Harlan had a macabre fantasy of the three of us sitting down in a graveyard with a videographer to record our reflections on life and death. Unfortunately, our schedules never quite harmonized, and Harlan's death put paid to his scheme.

Dan agreed to let me put two of his books into E-Reads. I happily discovered that I had reserved e-rights to *Song of Kali*, his dark and lurid horror novel set in Calcutta. The second, *Phases of Gravity*, had been published by a small press, and Dan granted E-Reads both print and e-book rights. *Phases* was about an astronaut whose voyage to the moon affected him so profoundly that it hampered his ability to form relationships on Earth. So vivid was Dan's description that I dreamed one night that I was his book's hero, standing on the moon and gazing at Earth. It was the most beautiful dream I ever had.

JOHN NORMAN

You may well imagine that Norman's saga of the fantasy world of Gor, a society in which males are the masters and females their willing bondservants, has been a source of intense controversy. Though the original twenty-five books in the series, published beginning in the 1960s, sold millions, fueled by Boris Vallejo's testosterone-stimulating cover illustrations, they went out of fashion as the women's movement rose up to reject them and feminist editors wouldn't touch them.

When I started E-Reads, I looked into their status and noticed something odd: a highly active cult of fans role-playing John Norman's characters in sims, fantasy communities with a social order, extensive rules and moral codes. A visit to the Gorean Campus on the Second Life website reveals a vast array of alluring (and expensive) Gor-themed fashions designed almost exclusively for women—women who do not seem to have gotten the memo that female submissiveness is flagrantly inconsonant with today's moral and political principles.

Inconsonant or not, I put Norman's books into the E-Reads program, and they became our bestselling series. The original twenty-five volumes have expanded to thirty-seven at this writing (the author, at ninety-two, is still cranking them out). But the great pity (from an agent's viewpoint at least) is that the author never trademarked "Gor" or the world he created. As a result, he doesn't receive a dime from the lucrative merchandise his work has spawned.

If you're intrigued, start with number one in the series, *Tarnsman of Gor.*

DAVE GROSSMAN

Sometime early in the '90s, Lt. Col. Dave Grossman, an Army officer and a West Point psychology professor, submitted a book to our agency that explored what he described as "the last unexplored intimate act," the act of killing another person.

Unlike that other familiar intimate act (the one you're thinking of), which has been explored exhaustively and publicly since the beginning of time, the emotions involved in taking another human life have seldom been examined in scholarly depth. For all but the hardened murderer, killing evokes feelings of horror, revulsion, guilt and shame. It is therefore hard to speak of publicly and, for many, impossible to speak of at all. Disturbing images exert dark and profound power over the psyche. The perpetrator is unable to banish them from memory but, at the same time, is unable to share them with those who might help them find relief.

This traumatic stress is of course particularly intense for soldiers.

Grossman spent years digging deeply and fearlessly into the subject, interviewing countless soldiers and veterans, psychologists and scholars, and drawing on the World War II observations made by Brigadier General S. L. A. Marshall. Out of Grossman's research came his groundbreaking book, *On Killing: The Psychological Cost of Learning to Kill in War and Society.*

What he learned was that most soldiers are instinctively loath to kill. Statistics from the American Civil War through World War II confirmed that only a small percentage of combatants fired their weapons, even in the face of attacking enemy. "Why did these men fail to fire?" Grossman asked in an article for *Greater Good*, a University of California Berkeley publication.

> As a historian, psychologist, and soldier, I examined this question and studied the process of killing in combat. I have realized that there was one major factor missing from the common understanding of this process, a factor that answers this question and more: the simple and

demonstrable fact that there is, within most people, an intense re-sistance to killing other people. A resistance so strong that, in many circumstances, soldiers on the battlefield will die before they can overcome it.[1]

How can the military function if its combatants can't pull the trigger? The answer is that armies have perfected indoctrination techniques designed to overcome the natural resistance of trainees to kill. "Soldiers are trained to think of themselves as killers and the enemy as deserving death," writes David Martin. "Killing is taken out of the abstract to become real and personal."[2]

But Grossman took this thesis a giant step further when he recognized that our violent media—movies, television, video games—are desensitizing children to violence the same way as soldiers. Even worse, they are teaching children the actual techniques of murdering, such as videogames that reward shots to the head. Out of Grossman's thesis came several books—*Assassination Generation* and *Stop Teaching Our Kids to Kill* (the latter coauthored with parent coach Gloria Degaetano)—urging our society to examine how the media have made it harder for children to distinguish between fictional violence and reality.

I placed *On Killing* with Little, Brown where it has sold over 750,000 print copies to date and is widely studied in military and police academies. We managed, however, to retain e-book rights, and when I launched E-Reads, Lt. Col. Grossman agreed to let us publish the electronic edition. It became the biggest-selling single book in our history.

ELIZABETH LYNN

The first generation of fantasy and science fiction writers of the American twentieth century was dominated by males who did not exactly put down the welcome mat of their clubhouse for female authors. Women characters were okay as long as they were sex objects or kickass warriors. The idea of romance in a science fiction novel of that time was as alien as creatures from another galaxy. Men were simply too busy conquering planets to conduct romantic relationships with women or harbor tender emotions.

Given that mindset, the idea of homosexual and lesbian sex and love was light years beyond the grasp of most fantasy and science fiction writers and readers of that era. Yet those were the groundbreaking themes of the works produced in the late 1970s and early '80s by Elizabeth A. Lynn. She won two World Fantasy Awards for them.

Decades later a genre called Science Fiction Romance arose that was short on science and long on eroticism. Worlds colliding gave way to Happily Ever After endings. They never came close to Elizabeth Lynn. Start with *A Different Light*.

On a personal note, Lizzie was a "godclient"—one of many—to my son Charles.

GREG BEAR

My representation of Greg Bear extended four decades back to the beginning of his career, which spans the evolution of the science fiction genre from mass market paperback originals to mainstream blockbusters to the E-Book Revolution. He fearlessly and presciently tackled cosmically massive subjects—nanomedicine, runaway genetic mutations, programming DNA, a future bureau of investigation, computer self-awareness, the existence of souls and much, much more. His storytelling techniques spanned an incredible range of narrative skills from the hard prose of space adventure to the dazzling poesy of a genetically transformed humanity. For this, he received numerous honors including five Nebulas, two Hugos and thirteen Locus Awards and served as President of the Science Fiction Writers of America from 1988 to 1990. He was instrumental in the genesis of the Science Fiction Museum and Hall of Fame in his home city of Seattle.

As an author, Greg thrived in stable, long-lasting editorial relationships. Unfortunately, the book business seldom reciprocated, as editors and even publishers came and went with little or no notice. As the business grew turbulent and the ground constantly shifted under his feet, he'd often preface a call to me with a querulous, "What the *hell* is going on in New York?!"

What indeed! As the business corporatized, then Hollywoodized, it became almost impossible to explain, even to this master of profundities. He could grasp the complexity of quantum computations, but Big Publishing remained a dark and unfathomable world to a humanist with a tender heart.

Like particles thrown off in an atom smasher, numerous books of his were tossed out of print by the industry's constant churning, and when I started E-

Reads, we identified more than twenty available for e-book publication, where they thrive to this day.

Greg died at seventy-one of complications of a heart attack, a lifetime cut short with so many fascinating ideas, projects, and complex themes waiting to be addressed. He was a writer of colossal vision who grasped the moral implications of science as few writers did or could. He was possessed of literary gifts that captured worlds in works of incomparable power and beauty that will endure forever.[3]

MARCO VASSI

To read Vassi's classic works of erotic fiction and his autobiographical masterpiece, *The Stoned Apocalypse*, is to realize that the man and the '60s were created out of the same fire and primordial elements. It is not, however, enough to say that he was a child of his age. It could just as accurately be said that the age was his fantasy, a fantasy so intense and compelling that it is impossible to read any of his books in one sitting: One must either jump into a cold shower or go for a long contemplative walk to reflect on the profundity of his insights into human sexuality.

As Fred Vassi, with no source of regular income, he tried his hand at what were then popularly known as sex novels, a genre of tame pornography that pandered to the fantasies of repressed males still mired in postwar inhibition (full disclosure: I wrote them, too). With the wide-eyed innocence and self-deprecating humor that characterized every venture he undertook, he showed them to me, his friend and a fledgling literary agent. He merely hoped to raise a few dollars with them. I told him that they were the most incredibly arousing works of erotic literature since Henry Miller and arranged for them to be brought out by Olympia Press, Miller's publisher, under the name Marco Vassi. Critics and reviewers confirmed my assessment. What distinguished his books from the rest of the pack was the application of his intelligence. He knew that the most erotic organ of all is the mind. He termed this fusion of mind and libido "Metasex."

For Vassi, the liberation of sexual emotions, paralleling the liberation of so many others in the late 1960s and early 1970s, promised a new age of beauty, love and honesty, and he lived his vision to the hilt—quite literally. For a long while it seemed to him impossible that this vision did not rest on the bedrock of reality. But the bloody hand of Vietnam and the corrupt fist of the Nixon presidency crushed the fragile beauty of the Flower Generation. The unbridled commercialism that became the 1980s captured and exploited the children of Woodstock, enriching half of them and killing the other half with sex, drugs and rock 'n' roll.

Finally, the horror of a new scourge, AIDS, visited death upon the bodies of those who had dreamed of eternal love, irresponsible fun and self-realization. It was then that Marco Vassi awoke from his dream of the '60s. When he did, the virus had entered his blood. The first malady of any consequence to come along—in his case, pneumonia—conquered his defenseless immune system and made short work of him.

E-Reads put most of his works into e-book format. Start with his first novel, *Mind Blower*, but his memoir, *The Stoned Apocalypse*, is also a must.

RICHARD S. PRATHER

Speaking of repressed males, I was among the tens of millions of readers who reveled in the bullet-paced adventures of Richard S. Prather's private eye, Shell Scott, capers densely populated with exquisite and willing females. Margalit Fox's *New York Times* obituary captured Prather perfectly. Prather's novels, she wrote, "were known for their swift violence, loopy humor and astonishing number of characters with no clothes on."[4]

Here I must issue the requisite disclaimer that the attitudes depicted in Prather's books include leering descriptions of pulchritudinous women. Men of the twenty-first century are too enlightened to smirk at such vulgar and distasteful depictions.

Prather's polar counterpart was Mickey Spillane, who told similar tales of mayhem and violence but devoid of much humor. Prather, by contrast, thrust his hero into rollicking dilemmas from which he had to extricate himself with self-deprecating wit and charm. A number of those situations involved dozens of barely clad (or completely unclad) females chasing after him. Here is an excerpt (from *Strip for Murder*) about a footrace in a nudist colony:

There was a monstrous, blood-chilling shriek from my left and I snapped my head around. I stared. A mountain of flesh was in motion, coming at me. At first it was a pink landslide, a volcanic lava flow of flesh rolling this way to inundate me; then I began to pick out individual segments of the mass: women. Women, all running as if each absolutely had to win, arms flying, legs pumping, everything doing something. It was appalling. Never in my wildest moments had I ever dreamed of anything remotely like this. There was a sound of thunder like buffalo stampeding; a trumpeting as of elephants stuck in swamps. Suddenly the race was over. I didn't have any idea who had won. All I knew was that three figures had streaked past me—and then I was drowning in naked babes.

As long as we're talking about it, start with *Strip for Murder*.

HARLAN ELLISON

There are a million stories about Harlan Ellison. Some of them are true. True or apocryphal, you could fill a book with them.[*] I'm afraid that if I start to scratch the surface of the legend, I will end up scratching to the center of the earth. So I will limit my memories to a single memorable account, a visit that Leslie and I and our seven-year-old son Charles paid to Harlan in Los Angeles as a stop on a California vacation. I had promised Charles that Harlan's home, Ellison Wonderland, held more marvels than Disneyland.

We were warmly welcomed by Harlan's lovely wife, Susan, and moments later the great man swept in, the glass of fashion in a shabby terrycloth robe and fuzzy slippers. He warmly greeted Charles—the Ellisons had no children of their own—and said, "Hey, kid, I have something you're going to love." He disappeared for a minute and returned with a box. He opened it and carefully removed the object inside. A Flash Gordon ray gun! "Mint condition!" he boasted, beaming. He demonstrated its operation to Charles, who took it in his hands and pulled the trigger. Sparks shot out of the barrel and it made a loud whiny VREEEE. Our son went trigger-happy—VREEEE! VREEEE! VREEEE! VREEEE!—zapping every alien in Ellison Wonderland.

---

[*] Somebody did: *A Lit Fuse: The Provocative Life of Harlan Ellison* by Nat Segaloff.

18. RESCUED

"Say thank you to Harlan," I told Charles, assuming it was a gift. But before he could open his mouth, Harlan snatched the toy out of his hands. "That's enough, kid." He returned the gun to its box and restored it to his collection of vintage memorabilia. I was very proud of Charles for not bursting into tears.

Many years later, as I have related, E-Reads rescued not only Harlan's books from calamity but his home as well.

Harlan once told me, "Put this on my tombstone: '*I mattered once.*'"

No, Harlan, you mattered then and you matter now and you will matter tomorrow, because your work will be available for eternity.

Any one of his collections will serve as an introduction, but I would start with *I Have No Mouth and I Must Scream*. To see the Great Man in action, go to YouTube and catch his classic rant, "Pay the Writer," excerpted from the documentary *Dreams with Sharp Teeth*, produced by Erik Nelson.

19. THE END

(2014)

Now hoist a mug of stout libation
As we commence our peroration:
May all your e-books be enhanced,
Your acquisitions well-financed.
And like a toddy mulled with schnapps
May every day be filled with apps.
Make sure you write upon your doormat
"Here we welcome every format."
And, like a ball festooned with spices,
A plethora of cool devices:
Kobo, Copia, Kindle, Nook,
Sony, iPad and... hmm, let's see, did I leave anything out?
Oh yes, I almost forgot . . . printed book.

T HE SALE OF E-READS AND DEPARTURE of its staff left our spacious office suite only half occupied, and we sought smaller quarters for just my agency. I found an ideal space just two blocks south. But as moving day approached and I surveyed our shelves and closets, I realized we were not just moving in space but in time as well, from the twentieth century to the twenty-first. Although we had upgraded our operation long ago, we had not completely cleared out all the artifacts of earlier days, perhaps out of sentimental attachment to an industry built on tangible materials, mechanical devices and analog communications. But the time had come to dispose of them.

19. THE END

Here, for instance, was a closet of submission boxes for book-length man-uscripts, and there, a shelf containing folders for mailing out short stories and articles. Another shelf held ring binders for archiving financial spreadsheets. In a utility closet stood the shopping cart in which we shlepped parcels to the post office. On the packing table, under a patina of dust, were a paper cutter, a three-hole paper puncher and a scale that measured cartons as heavy as thirty pounds for books shipped abroad or to Hollywood. In the words of W. S. Gilbert, "They'd none of them be missed." Or would they?

What did tug at my heart was the crank-operated pencil sharpener clamped like a lone sentry to a corner of a work table. We had vowed for years to buy an electric one, but there was something endearingly timeless about the *wugga-wugga-wugga* sound and the aroma of fresh-cut cedar. Dear sweet super-annuated little pencil sharpener! You deserve a gold watch for decades of loyal service! I couldn't remember the last time I'd used it. Indeed, I couldn't remem-ber the last time I'd used a pencil except as a stake to prop up a drooping plant on my windowsill.

I turned my gaze to the solid wall of file cabinets stuffed with contracts and royalty statements. We had long ago scanned and digitized the contracts. Royalty statements, now emailed to us or accessed on publishers' portals, were archived on our server and in the Cloud. Yet we were irrationally reluctant to dispose of all those paper documents, perhaps against the possibility that a cyberwar would wipe out the Internet and we'd be thrown back to the age of paper documents signed in ink. But we knew that when we moved, most of them would be shredded.

I felt a wave of nostalgia to behold the invalided veterans of the early elec-tronic era—clunky monochrome monitors, obsolete keyboards, primitive floppy-disk drives and generations-old servers long supplanted by sleek, com-pact models with capacities measured in terabytes. In a cardboard carton were some early models of e-book readers: a Franklin, a Rocket eBook, a Sony. I won-dered what we could get for these souvenirs on eBay, or perhaps we should save them to donate to some future e-book museum.

Sitting in a corner was the fax machine, unplugged after we realized that its sole function was to receive solicitations from loan sharks and cleaning ser-vices. And that obsolete phone system, relic of the twilight zone between the rotary and digital eras. Plus, the answering machine with a tinny greeting rec-orded by an employee who had moved on long ago. And—oh, yes!—the Ro-lodex, that catalog of failed lunch dates.

But it was the books—those lovely, handsome books—that elicited the biggest pang.

When I compared the wall dimensions of the old and new offices I realized with a jolt that I could not possibly fit all our books into the new space. Over decades, we had accumulated not just hundreds of titles but multiple copies of most of them, a dozen or more of each edition. How were we going to accommodate yards and yards of books that far exceeded the shelf space at our new location?

Leslie, my wife and business partner, came to the rescue. "Why," she asked, "do you need all those books?" The question was so dismaying it literally tied my tongue. "Why do I need all those books? This is *publishing*, for God's sake!" But then she added, "Or haven't you noticed the Digital Revolution, Mister Visionary?" She was right (as usual): Nobody requested printed books anymore. They asked for PDFs to be emailed to them.

As I had once been on the cutting edge of the Digital Revolution, I understood too well how our world has transformed from material to digital. But surrounded by walls and walls of books and all the poignant memories they held, I must confess I was in denial, deeply attached to paper and the world it represented. I was still a shameless print junkie. Still, we had no choice, and so—keeping no more than three or four of each title—we packed up the excess copies and shipped them to their authors or donated them to libraries. Then we got rid of all the other stuff, delivering the old electronics to recycle centers, shredding tons of paper and donating materials to one of those wonderful organizations that provide discarded office supplies to teachers. We put some objects out on the street, reasoning (correctly) that one person's trash is another person's treasure.

Then we called the movers and told them we were ready to move into the twenty-first century.[1]

20. BACK FROM THE FUTURE

(2023)

> *The upshot of this much-hyped tech?*
> *Bots hallucinating dreck,*
> *Automatons producing fiction,*
> *With clichéd plots and scrambled diction.*
> *The ballyhooed Large Language Model*
> *Spewing cockamamie twaddle.*
> *But who would not go into hock*
> *For one small tranche of Nvidia stock?*

F OR TWO YEARS AFTER THE ACQUISITION of E-Reads, I spent one morning every week in the bright, spacious offices of Open Road, in a tower on Maiden Lane with grand views of the lower harbor of Manhattan's East River. Their precincts were a beehive of young pioneers striving to achieve Jane Friedman's goal of acquiring ten thousand titles for her company (she eventually did). There I imparted such wisdom to her staff as I had garnered over the past fifteen years. These sessions eased my passage from agent-publisher back to my original status as an unhyphenated literary agent. This was somewhat disorienting at first, as if I had misplaced my alter ego, but by the end of my stint with Open Road, I had made the adjustment and was far enough removed from my sojourn in E-Land to step back and assess the publishing landscape with some objectivity.

Everything had changed.

I was looking at a trade book industry almost unrecognizably altered since the end of the twentieth century, almost all of it owing to digitization of the media. Jim Milliot, *Publishers Weekly*'s Editorial Director from the go-go '90s through 2023, summarized the metamorphosis in his valedictory editorial: "Technology has transformed publishing in every conceivable way, from how books are acquired to how they are printed, marketed, discovered, and sold."[1]

A twentieth-century Rip van Winkle awakening in the twenty-first would scarcely recognize the terrain or comprehend the jargon. I remembered the 1998 NIST conference, where techies and book people faced off like alien cultures striving to find a lingua franca. By the 2020s that problem was solved. Just about everyone in the book industry had upgraded to a higher technological mindset, from editors (using Track Changes software to edit scripts, for example) to contract managers (DocuSign) to bookkeepers (QuickBooks) to marketers, publicists, social media specialists, all the way down to mail clerks.

Automation of picking and packing books in warehouses, printing and drop-shipping PODs, tracking parcels in transit and countless other tasks had replaced the analog book business of the previous century, and it was all second nature to a generation that had grown up clicking, swiping, uploading, downloading, posting, texting, zooming, sharing, exploring and fearlessly experimenting. The community I had been part of was aging and yielding to a younger generation with its own style (wired), pace (lightning), language (tech) and culture (cool), a population that tossed around terms like "cache" and "source code" and "enhancements" and "file formats" and "metadata" as if they'd learned them at their mother's knee.

Many of my pre-Digital Era colleagues remained active, but clearly, the old regime was giving way to a new one that forged its own way of doing things. For instance, the new regime evinced a baffling aversion to telephone conversations and an almost paranoiac dread of phone negotiations. This was completely opposite from the garrulousness and conviviality of the previous generation. We loved to chatter, we loved to "hondle" (Yiddish for bargain) and we loved to tattle. The late Roberta Grossman of Kensington and I had a shtick: no matter how tied up she was, I could always get through to her by uttering these magic words to her secretary: "Tell Roberta, 'It's gossip.' " This communication invariably prompted Roberta to drop everything and jump on the phone. She took particular pleasure in trumping my bulletin with, "Ha! I knew *that* two days ago!"

The silence that had befallen our industry compelled me to write a guest editorial for *Publishers Weekly* entitled "Let's Make Some Noise!" "The vocal 20th century generation of publishing people," I wrote, "has begun to give way to a wired younger one that seems less comfortable and secure on the phone than on a keyboard." I concluded the article with this pathetic whimper: "Given the cone of silence in which some of us labor, a little stress-inducing noise and distraction would be welcome. Call me? Please?"[2]

I got dozens of supportive responses from colleagues—via email. Not a single phone call!

By 2023, the E-Book Revolution—in terms of both e-reading devices and the content they carried—had clearly run its course. Book books were back, and print was booming. In 2012, U.S. trade book sales had been $5.56 billion; ten years later, in 2022, revenues were up to $8.1 billion.[3] E-books' share of U.S. trade book sales, on the other hand, had dropped from nearly 22 percent in 2012 to 11 percent in 2022. (The Association of American Publishers doesn't calculate self-published sales.)

The concept of reading books on a dedicated device had lost its novelty. Adult ownership of e-readers declined from 32 percent in 2014 to 19 percent in 2023.[4] "E-books," wrote Milliot, "now augment print books, rather than replacing them as had once been widely prophesied."[5] That doesn't mean they weren't reading e-books, but if they were, it was not necessarily on Kindles and Nooks, but perhaps on iPads and tablets. Amazon does not disclose sales of the Kindle, but one source projected 900,000 units for 2023.[6] Compare that to the 60.4 *million* iPads shipped by Apple in 2022.[7]

Inexorably, tablets, smart phones and other interactive devices that offered eye- and ear-stimulating media—movies, music, videos, websites, games, audiobooks, emails, text messages, social media—lured attention away from the spartan act of reading an e- book, in black-and-white, with no recourse to all those fun and colorful distractions. Color played no small part in this shift. "The lack of color range in e-readers hindering their adoption is the major challenge for the growth of the global e-reader market," technavio.com reported, as if black print on white pages were a design flaw.[8]

Perhaps the most notable publishing trend was the shift from visual to aural, as audiobooks seized a significant share of reader attention. The number of titles in the U.S. market soared from fewer than 10,000 in 2011 to some 250,000 in 2023, generating over $1.8 billion in U.S. sales and $5.4 billion worldwide.[9]

Concomitant with audiobooks was the phenomenal growth in podcasting, with the number of listeners soaring by double digits from 333.2 million worldwide in 2020 to 464.7 million in 2023.[10] According to Backlinko.com, podcasts were a $23 billion industry in 2023 with a projected growth to $100 billion by 2030.

Just what this swing signifies is hard to say. Is reading over? Have we become a nation of listeners?

The answer seems more straightforward. Thanks to technology, such as the text-to-speech applications embedded in most word processing programs, it is easier than ever to adapt written text to spoken. Given that those programs are digital, a vast amount of material is available for conversion.[*] Audiobooks and podcasts satisfy people on the move or too busy or impatient to sit and read. The fact that many audiobooks and podcasts are free is no small attraction. And of course, audio is a vital necessity for the visually impaired.

It would be gratifying to say that writers also benefited from the myriad improvements that digital technology had bestowed on the rest of the industry. Unfortunately, though the tools of their profession had vastly improved, by almost every other measure authorship was in bad shape.

As I write this, in 2023, the disparity between have and have-not authors has dramatically widened. The fortunes of established stars have soared while the unbranded struggle to keep their heads above water. The mouth-watering six- and seven-digit deals headlined in *Publishers Weekly* and *Deal Lunch* vividly contrast with the plummeting earnings of garden variety professional writers.

A 2018 Authors Guild income survey, described by the Guild as its largest ever, found "incomes falling to historic lows to a median of $6,080 in 2017," down more than 50 percent from the $12,850 reported in a joint Guild and PEN survey taken in 2009.[11] Even after adding supplementary income from activities like speaking engagements and teaching, "respondents who identified themselves as *full-time* book authors still earned a median income of $20,300, well below the federal poverty line for a family of three or more." An October 2023 Guild survey concluded that "half of all full-time authors continue to earn below the federal minimum wage of just $7.25/hour from their books."[12] (Self-published and genre authors fared only slightly better.)

---

[*] Text-to-speech capabilities are now vastly enhanced by artificial intelligence, which can precisely reproduce the voices of professional narrators (and put them out of work).

Numerous factors account for dwindling author income and low morale, some of which I have already recounted. Piracy, the contraction of the marketplace, the collapse of mass market originals and the high share of e-book revenue (75 percent) taken by publishers have all contributed to make authorship a losing or discouraging proposition for too many writers. Even so, assaults on author earnings continue unabated.

The latest is a subscription model giving users, for a monthly fee, unlimited access to a retailer's complete library. But instead of basing author compensation on *books purchased*, this model calculates the royalty on *pages read* or, in the case of audiobooks, *minutes streamed*. The royalty comes to a fraction of a penny per minute. A recent royalty statement I reviewed showed that, for 498,854 minutes of listening (roughly 8,313 hours), the author earned $940.13, a fraction of what is paid for purchases of complete audiobooks through normal channels.[13]

Late in 2023, Spotify, the giant music aggregation platform, announced it was adding audiobooks to its vast library of offerings. Kim Scott, in a *New York Times* posting, sounded the alarm that this would accelerate the collapse of author income. Scott pointed out that

> in general, authors will be paid in full only if users *finish* the book. If someone listens to only a portion of the book, the author gets paid only for the amount of time the person listened. Given that many books are sold but never finished, many authors are likely to make significantly less under this model.
>
> Spotify is also likely to put a thumb on the scale of which audiobooks find an audience. It is, seemingly, pushing people to old, out-of-copyright books like, bizarrely, works of Karl Marx. Or it's pushing content from blockbuster artists with whom it has a relationship. Some listeners told me they were being pushed to the new Britney Spears autobiography, regardless of whether they were likely to be interested in Britney Spears. The middle class of book creators looks set to be even more squeezed, and revenue seems likely to be even more concentrated in the top 1 percent of authors.[14]

The gravest threat to the security of authorship (to say nothing of book publishing) is artificial intelligence. Granted, the potential for this technology is astounding and its prospective benefits unlimited, but in the microcosm of the literary and publishing professions, the potential for catastrophe is tremendous.

The underlying principle is that for AI to build a complete and reliable database, it must scan and collect all human knowledge. Much of it is in the public domain. A good deal of it, however, is proprietary and subject to copyright laws. There is no surefire way to exclude those works from the mass of information indiscriminately collected in the driftnets of AI trawlers. In fact, the aggregators have no desire to separate copyrighted works from the rest, because that would compromise the completeness of the database and distort the answers provided to users. Even if AI could find a way to separate copyrighted from noncopyrighted works, devising a formula to compensate owners would be a logistical nightmare. (In 2024, HarperCollins negotiated a licensing agreement with one company to permit some of their nonfiction titles to be used to train AI servers, splitting the proceeds between publisher and authors.)[15]

Therefore, as we have seen in the case of the Google Library Project, aggregators use a variety of dubious rationalizations to justify their inclusion of copyrighted works. Some invoke the fair use rule that entitles users to capture a sample or (to use a favorite word of infringers) "snippet"; some claim that the work is for education and research; some cite the benefits to society and improvement of the human condition; some say the AI application is "transformative," a legal term characterizing a new medium to which current regulations are not applicable. As we have seen, the judge who heard the Google case accepted these arguments, concluding that "all society benefits," and the Supreme Court declined to hear the appeal. And that is where the law stands as the courts take up AI infringement lawsuits.

Will authors and the Authors Guild fare better in their lawsuits than they did in the one against Google? The jury is still out. For example, In June, 2025, Judge William Alsup found that copyrighted books used by a company called Anthropic to train AI models without authors' consent was "exceedingly transformative and was a fair use." But Judge Alsup also noted that the company's downloading of copyrighted books (more than seven million of them!) from pirate sites may have been illegal.[16]

In February 2025, Thomson Reuters won a copyright infringement suit against an AI outfit, Ross Intelligence, which had used fair use as grounds for reproducing copyrighted materials from Thomson Reuters' Westlaw legal research company. The judge in the case declared "None of Ross's possible defenses hold water. I reject them all."[17]

These contradictory judgments are preliminary punches in a long heavyweight fight, but the odds against the Authors Guild and other plaintiffs remain daunting. Investment in generative AI is nothing short of astronomical, with company valuations in the tens or even hundreds of billions of dollars.[18]

One promising idea is for authors to compel publishers to create contractual language declaring that use of their text for AI training is a distinct right, like movie or audio, and authors can prohibit their publishers from granting those rights. A statement issued by Authors Guild in December 2024 proclaimed:

- AI Training Is Not Covered Under Standard Publishing Agreements
- Subsidiary Rights Do Not Include AI Rights
- Authors Retain Copyright
- Publishers Must Seek Author Permission
- Publisher Compensation Depends on AI Licensing Role
- Author Should Get Majority Share in AI Licensing Deals[19]

It can only be hoped that the courts will concur with the Guild's elegantly worded invitation to its members to sign a petition opposing AI's unauthorized use of their work:

> Generative AI technologies built on large language models owe their existence to our writings. These technologies mimic and regurgitate our language, stories, style, and ideas. Millions of copyrighted books, articles, essays, and poetry provide the "food" for AI systems. . . . You're spending billions of dollars to develop AI technology. It is only fair that you compensate us for using our writings, without which AI would be banal and extremely limited. . . . As a result of embedding our writings in your systems, generative AI threatens to damage our profession by flooding the market with mediocre, machine-written books, stories, and journalism based on our work. . . . The introduction of AI threatens to tip the scale to make it even more difficult, if not impossible, for writers—especially young writers and voices from under-represented communities—to earn a living from their profession.[20]

It's not just writers and books but civilized society that faces disarray. As Stephen Power points out, "What's at stake with the digital disruption of publishing isn't just the business of books but also the business of reading and thus the business of culture." The implications for the future of writing and publishing are deeply concerning.

In a survey of academic integrity in higher learning, Meazure Learning reported "just how rapidly student behavior is evolving with easier access to artificial intelligence. In 2025, nearly all students (92 percent) in a survey conducted by the Higher Education Policy Institute report using AI in some form—up from 66 percent in 2024. And 88 percent used generative AI tools to complete assessments, a staggering rise from 53 percent the previous year. These findings make clear that AI use is becoming nearly universal, and it's reshaping the academic integrity landscape at an unprecedented pace."

The next generation of novelists, journalists, historians and editors is coming out of those institutions of higher learning, where AI-facilitated research and writing is becoming unavoidably commonplace and where faculty and school administrators are all but helpless to stem the tide. It is hard to imagine why a student or scholar would ever again produce an original book report, term paper or research treatise when the touch of a key can do it for them. From there it's an easy jump to AI-assisted creation of literature.

Amid these gloomy forebodings, e-books cast a beam of hope. Works facing out-of-print doom were made accessible to new generations of readers and generated welcome income for authors. Even a few copies sold were heartening to authors. I remember a touching conversation with one writer whose latest E-Reads statement showed just one copy sold. "I've spent all morning wondering who that buyer was," she said wistfully.

Looking back at my fifteen-year odyssey, I take great pride in knowing I helped build a bridge between the analog and digital cultures and designed a business model that is still in use today. But the achievement of which I'm proudest is the rescue of those backlists.

The new paradigm had salutary effects on the management of my literary agency. How could it not, when every aspect of the business has been refined and streamlined? Formats, systems, contractual language and communications modes that didn't exist a decade or two earlier were now standard operating procedures. The technical expertise I had absorbed made me a more effective businessperson and author advocate.

One thing, however, did not change: the fundamental relationship between myself and the writers I represented. Imagination is a uniquely human trait, and artistic creativity is a treasure deserving the worshipful attention of those who minister to it. Art, music and literature come from some mysterious and inexplicable mental and spiritual wellspring. Despite the claims of AI proponents, they cannot be produced anywhere but the human mind.

But this gift also has to be monetized, for agents know a dark truth about writers: that they would give their work away for nothing if we didn't put a price on it. Helping authors buy the time to create is the highest calling for members of our profession. Indeed, managing creative people is itself a form of creativity. Authors need to be understood, cultivated, cherished, rewarded and protected. Above all they need to be read and remembered, to feel that they have left a footprint in the sands of time.

Years ago, in a flash of enlightenment, this grumpy old visionary glimpsed how that could be accomplished: just a simple portable lightbox on which books can be read.

ACKNOWLEDGMENTS

I SUBMITTED THE MANUSCRIPT OF THIS BOOK to publisher Karl Weber at 11:30 a.m. on a Wednesday in April. Less than 24 hours later, I received an email from him enthusiastically commending it and laying out his vision for its publication. By the end of that day we had shaken hands on terms. Best of all, he saw larger dimensions in it than I had dared imagine. Karl is that rarest of combinations, a master publisher, a gentleman and a mensch. I am totally indebted to him.

My appreciation to Sarah Yake for her passionate advocacy and unwavering friendship.

I am beholden to editor Stephen Power for his perceptive and thought-provoking notes on my early draft.

And thank you, Lauren King, for designing my website and patiently guiding me through the complexities of social media.

Arthur Maisel demonstrated that copyediting is an art form, and I am incredibly thankful for his meticulous and discerning job.

I have been buoyed by the loving support of my son Charles and daughter-in-law Erica and their wise counsel and feedback. And a special shoutout to Erica's parents, my publishing buddy Janet Parker and accountant Phil Parker, with a big high-five to Phil for rescuing me from a mortifying calculation error.

Un fuerte abrazo to Liced Cintron, my loyal friend of many decades, faithful manager of my businesses and relentless hunter of bank statement discrepancies.

A special thank-you to Stefan Rudnicki for his unforgettable kindness.

Beyond all else is my incalculable gratitude to my wife Leslie Tonner Curtis, stalwart companion on both the journey and the book. Thank you, Leslie, and thank you again.

Richard Curtis

APPENDIX A

STANDARD E-READS® PUBLICATION CONTRACT
WITH COMMENTARY BY THE AUTHOR

Contract for Publication

TITLE OF CONTENT

Date

Distribution and License Agreement between E-Rights/E-Reads, Ltd., 171 East 74th Street, Suite #2, New York, NY 10021, ("E-Reads") and

Author Name
Author Address

(hereinafter referred to as "Content Provider") is the owner or authorized licensee of a literary work or other materials listed in Exhibit A (hereinafter referred to as "Content"; all titles listed in Exhibit A are a material part of this agreement).

1. Grant of rights. Content Provider grants to E-Reads the exclusive right, in the English language throughout the world, to distribute Content in electronic and print on paper format and to display, distribute and advertise or cause to be displayed, distributed and advertised all digital and printed versions of Content granted by Content Provider to E-Reads online and through such other distribution and sales channels as E-Reads may elect. For purposes of this agreement "electronic" shall mean a digital version of the some or all of the text of the Content that is made accessible to end users on a controlled basis as a discreet product via (i) storage on a physical medium, such as a CD-ROM, that

> We created this contract *ex nihilo*—there was no precedent to follow.

> Unlike traditional print deals, digital deals are always worldwide as it's impossible to prevent e-books from crossing borders. We reserved translation rights to the author, since the market for these was still embryonic.

The contract mentions CD-ROMs, but we never produced any.

is distributed to the end user; or (ii) downloading by the end user from the Internet or other computer network or server for reproduction and reading on a computer or similar device; or (iii) the end user's accessing a file stored on a remote computer or server that contains some or all of the text of the Content. Such electronic versions may contain devices to facilitate reading and use of the text, such as hyperlinks. From time to time, Content Provider and E-Reads may mutually agree to add other content to this agreement. Such additional content shall be expressly incorporated into Exhibit A and shall be bound by the terms of this agreement. The term granted by E-Reads for such additional content shall commence as of the date of an Additional Content Order signed by Content Provider.

The parties agree to the following terms and conditions.

Traditional print royalties ranged from 6% to 15% of list price. I considered e-book sales a form of reprint, for which a 50/50 revenue split was standard. Heavy discounting would have made royalties based on list price impractical, so we based our royalties on whatever we actually received. Our physical books were produced by print on demand, which I regarded as a tangible form of e-books, so a 50% royalty for these was appropriate.

(a) Payment for Content downloaded to personal computer or electronic reader: E-Reads agrees to pay Content Provider Fifty percent (50%) of any net revenues received by E-Reads earned from distribution of Content on a per-transaction basis. "Net revenues" in this context are defined as advertised list price of Content less retailer, distributor or other third party discounts, fees, commissions, and taxes if any.

(b) Payment for Content printed on paper, bound, and shipped to wholesalers, retailers, consumers or other purchasers: E-Reads agrees to pay Content Provider Fifty percent (50%) of any net revenues received by E-Reads earned from distribution of printed Content. "Net revenues" in this context are defined as advertised list price of Content less retailer, distributor or other third party discounts, fees, commissions, and sales taxes if any and less the cost of manufacture, handling, postage, freight charges, and taxes if any, but in no event shall net revenues reported to Content Provider be a negative number.

(c) Content Provider grants to E-Reads the right to co-publish Content with or distribute Content through a third party or to license trade hardcover or paperback book and book club publication rights in the English language, and E-Reads agrees to pay Content Provider eighty-five percent (85%) of any net revenues received by E-Reads from such license.

(d) Audio recording and broadcasting rights to Content, as commonly defined in the audio industry and including but not limited to CD, audiotape, and other voice recording, are expressly reserved to Content Provider. However, Content Provider grants to E-Reads the right to embed text-to-speech programs in its electronic editions of Content and to podcast or otherwise enable third parties to activate such text-to-speech programs.

2. *Production Fee.* To defray the cost of producing the Content, E-Reads shall charge a production fee of $250.00, per title, recoverable from royalties payable to Content Provider.

3. *Royalty statements.* E-Reads shall issue to Content Provider a report of earnings as of March 31, June 30, September 30, and December 31 of each calendar year and remit report and payment if due not later than sixty (60) days following said dates. No earnings shall be payable on free or promotional distributions. Any net revenues (as defined above) earned by E-Reads in the capacity of retailer shall be included in accountings rendered to Content Provider. In the event that this Agreement and/or any Additional Content Order covers more than one title, E-Reads shall not jointly account royalties due to Content Provider.

Content Provider or Content Provider's duly authorized representative shall have the right, upon reasonable notice during usual business hours but not more than once each year, to examine the records of E-Reads at the place where the same are regularly maintained insofar as they relate to the Content.

4. *Display and distribution of Content.*

(a) Within sixty (60) days of execution of this agreement Content Provider agrees to provide, Content Provider shall furnish, at Content Provider's expense, a proofread digital text file of Content to E-Reads in accordance with format guidelines defined in Exhibit B in digital format acceptable to E-Reads. In the event that Content Provider wishes to revise Content after E-Reads has commenced production or released Content, Content Provider shall pay to E-Reads the cost of reformatting and re-releasing Content. If customer complaints about editorial

We never produced audios. Legal rules regarding impaired persons meant we had to permit distributors to convert our texts into audios, which led to tension with genuine audiobook publishers.

Asking authors to subsidize publication was a radical innovation, justified by the big 50% royalty. We laid out the fee and recouped it from royalties, so authors felt no pain.

Traditional print publishers issued royalty statements just twice a year.

errors in the published Content caused by Content Provider require E-Reads in its sole judgment to withdraw Content from sale and distribution in order to proofread, correct, reformat and rerelease Content, the cost of such proofreading, correction, reformatting and rerelease shall be charged to Content Provider and recovered from royalties payable to Content Provider.

(b) If Content Provider is unable to provide a proofread digital text file of Content to E-Reads, Content Provider shall provide to E-Reads at Content Provider's expense one copy of the original printed edition. E-Reads shall convert Content into digital format at E-Reads' expense and author shall proofread the converted file. E-Reads shall produce computer-readable format within 180 days of delivery to E-Reads of hard copy version of Content.

(c) Content Provider shall furnish to E-Reads where feasible, at Content Provider's expense, copies of reviews, publicity releases, author biographical sketches, author photo with appropriate credit, a link to Content Provider's web site, and other material to assist E-Reads to display and publicize Content.

(d) E-Reads agrees to place the Content online for distribution as soon as technologically feasible but no later than 90 days after Content acceptably formatted for distribution has been delivered to or completed by E-Reads.

(e) Content Provider permits E-Reads to display Content or any portion of it online or in hard copy at no charge to customers for purposes of advertisement and promotion, provided E-Reads takes reasonable measures to protect the copyright in the Content.

(f) Content Provider hereby grants E-Reads the right to edit Content for technological compatibility with mode of distribution. However, E-Reads shall not otherwise edit Content without Content Provider's express approval.

5. Grant of License. Content Provider hereby grants an exclusive license to E-Reads to distribute the Content for ___ years from the date that E-Reads commences distribution of Content. At the end of the term, this Agreement shall automatically be extended on a year to year basis as long as E-Reads pays to Content Provider an extension fee or royalties (or a combination of extension fee and royalties) totaling no less than $100.00 commencing the final year of the original term. If extension fee and/or revenues paid by E-Reads at the end of the final year or at the end of any subsequent extension year do not total $100.00, and in any event at any time following the end of ten years from the date that E-Reads commences distribution of Content, Content Provider may terminate this Agreement by serving to E-Reads written, sixty-day notice. Upon termination of this agreement E-Reads shall return to Content Provider the RTF version of Content but shall retain ownership of the formatted e-book and print files. Content Provider may purchase formatted files from E-Reads for the original production cost incurred by E-Reads.

If the term of copyright of Content expires prior to the end of the term of this agreement, Content Provider shall give written notice to E-Reads of such expiration.

6. Warranties. Content Provider warrants that he/she is sole author of the Content; that he/she is the legal proprietor of all rights in the Content; that he/she has the power, right, and authority to enter into this Agreement, free and clear of any lien, claim, or debt, and that the grant will not impair the rights of any third parties and will not be libelous or in any way illegal; that all content provided by Content Provider will not infringe on the copyrights, trademarks, service marks, patents, or other intellectual property or personal rights held by any third party. E-Reads warrants that it will perform its services in a competent and professional manner, has taken every reasonable measure to protect Content from unauthorized reproduction and to ensure that the technological aspects of the product operate substantially according to the industry standards. E-Reads does not warrant that it will be able to correct all reported defects or that use of the web sites or links to or from the site will be uninterrupted or error free. E-Reads makes no warranty regarding features or services provided by third parties that are provided "as is" and "as available." Content Provider acknowledges and agrees that except for the express warranties provided in this

Since almost all of the books we acquired had been previously published, it was *imperative* that we had a written notice from the original publisher reverting rights to the author.

We hated pirates—and our contract obligated us to fight against them valiantly.

Agreement, all warranties whether express, implied, or statutory, and all obligations and representations as to performance including all warranties of merchantability or fitness for a particular purpose are hereby expressly excluded and disclaimed by E-Rights/E-Reads, Ltd. Content Provider acknowledges and agrees that in no event shall E-Rights/ E-Reads, Ltd. or its respective directors, officers, employees, partners, affiliates, or agents be liable for special, incidental, exemplary, consequential or indirect damages or for the loss of anticipated profits to Content Provider or its customers or any other person under any provision of this Agreement or any accompanying Additional Content Order.

7. Nonperformance. In the event E-Reads fails to furnish Content Provider with a royalty statement or to remit royalty due in any given reporting period, and does not cure such failure within thirty (30) days of receipt of written notice from Content Provider via United States postal mail to above address, Content Provider may cancel this Agreement by serving notice in writing to E-Rights. Such cancellation shall not mitigate E-Reads' obligation with respect to royalties then due.

8. Acts of God. Content Provider agrees that E-Reads shall not be liable for damages of any kind arising from Acts of God or any condition beyond its control.

9. Indemnification. Content Provider agrees to defend, indemnify, and hold harmless E-Reads and its respective directors, officers, technology partners, employees, affiliates, and agents from all claims, actions, losses, liability, damages, costs, and expenses including reasonable attorney's fees and expenses arising from Content Provider's warranties or provisions or actions under this Agreement. Content Provider shall indemnify and hold harmless E-Reads against liabilities that arise from (1) the materials, work or content provided; and (2) any content provided that infringes on the intellectual property or other rights of a third party or which is found defamatory. Each party agrees to promptly notify the other in writing of any indemnifiable claim pertinent to this provision and to give the other party the opportunity to defend or negotiate a settlement of any such claim, and to cooperate fully in defending such claim. E-Reads reserves the right at its own expense to

> We never missed a single royalty payment.

> Thanks to my fanatical conservatism (or timidity), we were never sued by anyone nor ever served with any legal claims.

> Since most of our books had been previously published by major houses, author breaches of warranties were basically no issue.

assume its exclusive defense of any matter otherwise subject to indemnification. E-Reads shall not settle any claim with respect to Content
Provider without Content Provider's written approval.

10. Reserved Rights. All rights not granted hereunder are reserved to
Content Provider including but not limited to print and electronic book
publication in languages other than English; spoken and performed audio other than embedded text-to-speech audio programs; and theatrical
motion picture, television, animated, stage, radio, and all other dramatic adaptations. No license of any reserved rights not expressly
granted by Content Provider to E-Reads hereunder shall be made by
E-Reads without the prior approval of Content Provider. Terms for acquisition of such reserved rights shall be a matter of separate agreement
on terms to be negotiated.

> We did not take any right that we could not exploit.

11. Bankruptcy. In the event that E-Reads files for or is placed in bankruptcy under Article 7 of the United States Bankruptcy Code, all rights
granted to E-Reads hereunder shall immediately revert to Content Provider.

12. Assignment. This Agreement may not be assigned by either party
without the written consent of the other, except to a company which
succeeds to all or substantially all of the business or assets and under-
takes to perform the obligations, of such party.

> This clause took effect in 2014 when Open Road acquired E-Reads.

13. Miscellaneous. The Agreement consists of this Distribution and
License Agreement and any and all Additional Content Orders, and
this Agreement supersedes all prior proposals and understandings, oral
or written and is the entire understanding between the parties. This
Agreement binds and inures to the benefit of each party's permitted
successors and assigns. This Agreement may be modified only in writing executed by both parties, provided that upon execution of additional content orders the terms of this agreement remain in full effect.
The laws of the State of New York shall govern this Agreement.

14. Reversion. In the event E-Reads has not completed production or
commenced distribution of Content in accordance with 4(b) or 4(d)
above, Content Provider may cancel this Agreement with no further

obligation of the parties to each other. In the event of such cancellation, Content Provider may retain any advance paid by E-Reads and shall not be responsible for production or proofreading costs if any incurred by E-Reads.

15. Author purchase of copies of printed book. Content Provider may purchase copies of the print version of Content from E-Reads at 50% discount off the list price plus shipping, handling, and taxes if any, for Content Provider's personal use or for resale outside of book trade wholesale or retail channels. No royalty shall be payable to Content Provider for book purchases made under the terms of this paragraph. I have read the terms of this Agreement, understand my duties and obligations, and agree to perform in good faith under these terms and conditions.

Content Provider

_____

Print Name

_____

On behalf of (company name if any)

_____

Signature and date

_____

Social Security or Federal Tax ID Number

_____

On behalf of E-Rights/E-Reads, Ltd.

_____

Signature and date

E-Rights/E-Reads ("E-Reads") Limited Release

Richard Curtis Associates, Inc. (RCA Inc.), a company owned by Richard Curtis, currently represents me on the sale of publication rights to my books. It is my desire to enter into a separate agreement with E-Rights, a distinct and separate company in which Richard Curtis has an ownership interest. I acknowledge that no conflict of interest exists between E-Reads and Richard Curtis, RCA Inc., its officers, directors and employees.

RCA Inc. has agreed to waive all commissions on moneys payable to me by E-Rights or any subsidiary or assignee of E-Reads.

This provision established our policy of no conflict of interest between our agency and our publishing company.

I further acknowledge that all my questions have been answered to my complete satisfaction with regard to E-Reads and that I have had an opportunity to consult with an attorney or other representative of my own choice.

As such, I acknowledge Richard Curtis's ownership disclosure and hereby accept RCA Inc.'s waiver of commission on revenues due to me from E-Reads and I further waive the right to outside counsel or representation with respect to the E-Rights/E-Reads, Ltd. Distribution and License Agreement dated January 10, 2006 including any subsequent Additional Content Orders. I hereby agree that my interests are fully protected by such Agreement and by this Release, and I hereby release Richard Curtis, RCA Inc., its officers, directors, and employees from any claim of liability or damage arising from any conflict, actual or perceived, arising from RCA Inc.'s representation of my publication rights and Richard Curtis's ownership interest in E-Rights/E-Reads, Ltd.

We offered all authors, including Curtis Agency clients, the opportunity to be represented by a lawyer or another agent in dealing with us. Not a single Curtis Agency client ever took advantage of this resource.

The laws of the state of New York shall govern this Release. Nothing herein shall limit or alter the warranties, terms, conditions, and provisions of any E-Rights Distribution Agreement or Additional Content Order that I sign.

Accepted

_____

Signature and date

Content Format and Proofreading Requirements

This instruction sheet offered authors the opportunity to be involved in the editorial process from the start. Some who accepted were humbled by the amount of work it took to produce a clean, attractive e-book.

E-Reads would like to thank you for offering to take the time to proofread your book. As you can well imagine, the errors that the scanning process introduces are not always easy for an impersonal third-party proofreader to catch. By going over the scanned text yourself, you have the opportunity to catch errors, amend phrasing, and even add new passages. In the end, you will have a manuscript archived in digital form that will be of invaluable service to everyone here at E-Reads.

However, please take extra special care in ensuring that every word and punctuation mark is correct. Once the text has been submitted to us as a master copy for e-book and print editions, you may be charged for any subsequent proofreading that you overlooked, or any editing you wish to make in the future.

Using a word processor such as Word, Word Perfect, or similar, will allow you to format your text as a Rich Text Format file (commonly known as RTF), which allows you to indicate Page Breaks and specify font family attributes like Bold and Italics. When you save your text, be sure to "Save As RTF."

Finally, when proofreading your text, please follow the guidelines below to help you set your text for our conversion process.

Properly Formatting RTF Files

1. Remove unnecessary text.
   a. Be sure to remove all front matter material (except the title, original copyright holder and date) that pertains to the original publisher's edition.
   b. Also, carefully remove all <image= " "> tags.

2. Insert PAGE BREAKS before every section and chapter (see your word processor's documentation, but might be indicated as "^m").

3. Scene Breaks must be indicated by a series of asterisks and not by a double spaced line. Our standard format uses a single line break

followed by 3 asterisks with one space between each (centered on the page), followed by another line break. Special double spaced scene breaks can replace asterisks in the print editions or in special circumstances, but please request this change by specifying such locations in a separate list saved as an RTF file, so that we can apply changes ourselves.

This is how our scene breaks should look:

* * *

4. E-Book formats typically have difficulty with **Bold** and *Italics* used at the same *time*. Usually it is best to choose bold over italics. Please do not use both at the same time.

5. Any photos or graphics previously included must be saved separately as unique files, preferably JPG at 150dpi, and listed with desired locations in another RTF document.

6. Please include a final word count at the top of the document. You must include attribution and copyright release information for any such work you include.

7. If possible, please submit jacket copy (one paragraph blurb/synopsis that entices readers) with the proofed RTF either clearly marked at the top of the document or as a separate RTF attachment. Your previously collected reviews are also welcome.

8. E-Mail or send your files on disc to E-Reads and retain a copy for yourself.

APPENDIX B

SAMPLE E-READS NEWSLETTER

July 31, 2000

Dear Authors (and other content providers):

I'm enclosing herewith E-Reads' second quarter 2000 royalty statement. For those of you whose production charge has been recouped by royalties, a check is also enclosed.

I want to bring you up to date on our progress on a number of fronts and assure you that however modest these early statements may be, the future is very rosy and we're optimistic that within the year your royalties will increase substantially. The digital revolution has finally taken hold in the book industry and your early commitment to e-reads has given you a significant head start over authors who are just waking up to the possibilities.

*Our web site. We are on the last leg of installation. As many of you know, we have delayed launch until we found PC download software that we felt was fast, reliable, and, most important of all, state-of-the-art secure. We are satisfied that the Glassbook Reader satisfies these demands. Readers visiting our site for the first time who wish to download an e-reads title will first download the Reader free of charge. Thereafter, they may order any title they wish and it will be readable on their installed Glassbook Reader. Reproduction of the text by third parties will be restricted, and though we all know that any determined hacker can pirate a book, we also know the best way to combat piracy is to make books available at reasonable prices via reputable channels.

Once our web site is up and running we expect the number of sales, and the royalty per sale, to increase sharply, because we will not only be sharing our publishing royalty with you but our retail royalty as well. Your books will be carried on other retail sites, multiplying your sales. In addition, we are developing an affiliate program that will enable other online sites to list our titles. We will do the fulfillment on our site and will pay commissions to our affiliates for sales of your books that they make. The best news is that we will be making you an affiliate, meaning that by selling your own titles on your web site, you will be entitled to a bonus over the percentage provided for in your e-reads agreement. We'll be in touch when this feature is in place.

*Rocket eBook and SoftBook. As you may have heard, the parent companies of these two pioneers in dedicated handheld readers were acquired over the winter by Gemstar, the company that owns TV Guide and the cable channel that scrolls your local television schedules. Gemstar discontinued the manufacture of the first version of the Rocket eBook and SoftBook and will soon release versions that are lighter, more functional, have improved screens (the SoftBook's will be in color) and are otherwise vastly improved. Gemstar is planning a huge television promotion in the coming Christmas season. The number of devices in circulation at the end of 1999 was approximately 10-15,000; Gemstar's goal for the Christmas season is half a million. Sales of our titles via these devices will grow concomitantly. As the saying goes, a rising tide lifts all boats.

*Palm Pilot. All of our titles are being converted for display on Palm Pilot, and although the current models are not the most desirable means for reading a book, with over 7 million Palm Pilots out there, the potential for growth in this area is huge.

*Microsoft has jumped into the e-book game with its Microsoft Reader. In principle this is like the Glassbook Reader but operates on a different system. The Reader's launch is scheduled for August and all of our titles are being converted for this application. Knowing that Microsoft doesn't do things by halves, we are optimistic that sales of our titles on their Reader will generate big royalties.

*Print. Most of our books are available for production in traditional bound paper format. We have a license with LightningSource (formerly Lightning Print) to make those books available in that format. As Lightning is owned by Ingram, the world's largest book distributor, those books automatically go into the Ingram database and can be ordered by all retailers ranging from Amazon.com to the independent book shop around the corner. In addition, we have entered into a strategic partnership with LPC, a book distributor that has agreed to do traditional print runs for those books on our list that have unusual sales potential. LPC has created a special imprint just for us, Olmstead/e-reads, and many of our titles are on sale or in production including reissues of Janet Dailey's bestselling HEIRESS, Harlan Ellison's anniversary edition of *Deathbird Stories*, and Graham Masterton's horror classic *The Manitou*.

*Bookface.com. This exciting site is carrying a number of our titles on an experimental basis. Books are displayed on the Bookface site, and revenue is generated from advertising on the site. Browsers may order the book at any time. Bookface offers an intriguing alternative to the traditional bookselling model and are hopeful it will generate revenue for our authors.

*Royalty reporting. We are putting the finishing touches on a state of the art digital royalty reporting program. I don't want to say more until we've test-driven it, but I'm confident it will leave traditional royalty reporting systems in the dust. I'm hoping you'll see it in our next reporting period at the end of October.

*Original books. Although we founded e-reads on previously published books, of which we have close to 1000 under contract, we are about to publish our first original titles, and are acquiring more. We hope to have a fully operational editorial department in place within one year, focusing on our stock in trade, category fiction, as well as general fiction and nonfiction. Unlike our competitors we will not become a vanity publisher, and we believe that in the last analysis, readers will select book sites carrying titles of proven quality.

*Visibility. Rather than rush our launch, we have spent the last year learning and perfecting publishing skills, creating strategic relationships with major companies in the publishing and e-book field, and acquiring an impressive list of titles. As soon as we are online we will move to the next phase of development—marketing and promoting your books and making e-reads a household name.

Thank you for your support and patience. I am confident that these drops in your royalty bucket will swell as each of the above elements matures.

Sincerely,

[signed] RICHARD CURTIS

SOURCE NOTES

PRELUDE: APPOINTMENT IN GAITHERSBURG

[1] Size of auditorium confirmed in an email to the author dated October 11, 2023, from Karen Startsman, Public Affairs Specialist, NIST Conference Program.

[2] "The Flowering of the Book Trade," *Publishers Weekly,* March 30, 1998, https://www.publishersweekly.com/pw/print/19980330/39933-pw-the-flowering-of-the-book-tech-expo.html.

[3] Proceedings of NIST-IR 6372 Electronic Book '98 Workshop, https://tsapps.nist.gov/publication/get_pdf.cfm?pub_id=151439.

[4] William M. Bulkeley and Alex Pevzner, "Russ Wilcox Steps Down at E-Ink—Smart Energy Venture Next? Xconomy," *Wall Street Journal,* June 2, 2009, https://www.wsj.com/articles/SB124387146323172505; Kit Eaton, "E-Ink's Sale Clears Path for Color Printer in 2010," *Fast Company,* June 1, 2009, https://www.fastcompany.com/1288671/e-inks-sale-clears-path-color-kindle-2010.

[5] "Christina Lampe-Önnerud," *Wikipedia,* https://en.wikipedia.org/wiki/Christina_Lampe-%C3%96nnerud.

[6] Heath Evans, "Content Is King: Essay by Bill Gates, 1996, *Medium,* January 29, 2017, https://medium.com/@HeathEvans/content-is-king-essay-by-bill-gates-1996-df74552f80d9.

1. ANALOG AGENT

[1] Bob Brown, *The Readies* (New York: Roving Eye Press, 2014), https://monoskop.org/images/e/e8/Brown_Bob_The_Readies.pdf.

[2] Edwin Mc Dowell, "Publishing: Agent Says Writers Are Cheated," *New York Times,* February 4, 1983, https://www.nytimes.com/1983/02/04/books/publishing-agent-says-writers-are-cheated.html?searchResultPosition=8; Angela Hoy, "Stop

Gluttonous Purchasing Practices by Bookstores! Why ALL Books Should Be Non-Returnable," *WritersWeekly.com*, January 7, 2021, https://writersweekly.com/angela-desk/stop-gluttonous-purchasing-practices-by-bookstores-why-all-books-should-be-non-returnable-by-angela-hoy-booklocker-com-writersweekly-com-abuzz-press-and-pubpreppers-com.

[3] Lynn Neary, "Publishers Push for New Rules on Unsold Books," *Morning Edition*, NPR, June 13, 2008, https://www.npr.org/2008/06/13/91461568/publishers-push-for-new-rules-on-unsold-books.

2. THE DREAM OF PORTABILITY

[1] "Timeline of Computer History," Computer History Museum website, https://www.computerhistory.org/timeline/memory-storage/.

[2] Peter Yianilos, Proximity Lab website, http://pnylab.com/products/about/main.html.

[3] Cade Metz, "Michael Hawley, Programmer, Professor and Pianist, Dies at 58," *New York Times*, June 24, 2020, https://www.nytimes.com/2020/06/24/technology/michael-hawley-dead.html.

[4] "Franklin Electronic Publishers," *Wikipedia*, https://en.wikipedia.org/wiki/Franklin_Electronic_Publishers.

[5] Associated Press, "E-Publisher Wins a Big Round," *Wired*, March 13, 20902, https://www.wired.com/2002/03/e-publisher-wins-a-big-round/; Hillel Italie, "Random House Settles with E-Book Seller," *Edwardsville Intelligencer*, December 3, 2002, https://www.theintelligencer.com/news/article/Random-House-Settles-With-E-Book-Seller-10525948.php.

[6] Random House, Inc., v. Rosetta Books Llc, 283 F.3d 490 (2d Cir. 2002), https://law.justia.com/cases/federal/appellate-courts/F3/283/490/484596/.

[7] "Harper Collins Makes a Costly Stand on E-Book Royalties in Open Road Litigation," Authors Guild website, November 11, 2014, https://www.authorsguild.org/industry-advocacy/harpercollins-makes-a-costly-stand-on-e-book-royalties-in-open-road-litigation/.

[8] "History of Video Games," *Wikipedia*, https://en.wikipedia.org/wiki/History_of_video_games; Benj Edwards, "Happy 30th B-Day, Gameboy:Here are six reasons why you're #1," *Ars Technica*, April 21, 2019, https://arstechnica.com/gaming/2019/04/game-boy-20th-anniversary/.

[9] Robert Coover, "The End of Books," *New York Times*, June 21, 1992, https://www.nytimes.com/1992/06/21/books/the-end-of-books.html.

[10] Andrew Liptak, "J.D. Salinger's Catcher in the Rye will be published as an ebook for the first time," *The Verge*, August 11, 2019, https://www.theverge.com/2019/8/11/20801082/catcher-and-the-rye-j-d-salinger-ebook-first-time.

4. THE AGE OF MIRACLES

[1] "Watchman FD210BE," Radiomuseum website, https://www.radiomuseum.org/r/sony_watchman_fd210befd_210_b.html.

[2] Tamsin Oxford, "Getting connected: A history of modems," Tech Radar website, December 26, 2009, https://www.techradar.com/news/internet/getting-connected-a-history-of-modems-657479.

[3] Wesley Chai, "What is Ethernet?" TechTarget website, October 26, 2023, https://www.techtarget.com/searchnetworking/definition/Ethernet.

[4] "Tim Berners-Lee, Robert Cailliau, And Invention of The World Wide Web," LivingInternet.com, https://livinginternet.com/w/wi_lee.htm.

[5] "The History Of Ecommerce: How Did It All Begin?" Miva website, November 23, 2020, https://blog.miva.com/the-history-of-ecommerce-how-did-it-all-begin.

[6] NJ, "How Many Websites Are There in the World?" Siteefy website, March 20, 2025, https://siteefy.com/how-many-websites-are-there/.

[7] "Hypertext," New World Encyclopedia, https://www.newworldencyclopedia.org/entry/Hypertext.

[8] "14 iconic websites that show off classic 90s web design," Webflow, https://webflow.com/blog/90s-website-design.

[9] Amanda Zantal-Wiener, "A Brief Timeline of the History of Blogging," HubSpot, October 20, 2020, https://blog.hubspot.com/marketing/history-of-blogging.

[10] "About: RankDex" Archived 2015-05-25 at the Wayback Machine, RankDex; accessed 3 May 2014. Cited in Wikipedia RankDex, https://web.archive.org/web/20150525015816/http:/www.rankdex.com/about.html.

[11] Jim Taylor, *DVD Demystified* (New York: McGraw-Hill, 1998) 1st edition, p. 405.

[12] *Statistical Yearbook: Cinema, Television, Video, and New Media in Europe* (Bloomington, IN: Indiana University, 1996), accessed on Google Books, https://books.google.com/books?id=MfoHAQAAMAAJ&focus=searchwithinvolume&q=%22should+reach+USD+30+billion+by+1998%22.

[13] Michelle Kratz, *Corporate Influence: How the Media Merger Trend Changed the Book Publishing Industry and the Distribution of Information* (New York: Pace University, Master's Dissertation, 2009), https://digitalcommons.pace.edu/cgi/viewcontent.cgi?article=1018&context=dyson_mspublishing.

[14] Kratz, *Corporate Influence.*

[15] Adam Hayes, "Long Tail: Definition as a Business Strategy and How It Works," Investopedia, January 26, 2025, https://www.investopedia.com/terms/l/long-tail.asp.

[16] "Amazon is founded by Jeff Bezos," One Day in History website, March 2, 2025, https://www.history.com/this-day-in-history/amazon-opens-for-business.

[17] Barnes & Noble 1998 Annual Report, https://www.annualreports.com/HostedData/AnnualReportArchive/b/NYSE_BKS_1998.pdf.

5. TURNING POINT

[1] Ake Grönlund et. al., editors, *Electronic Government and the Information Systems Perspective: Second International Conference, EGOVIS 2011, Toulouse, France, August 29-September 2, 2011, Proceedings* (New York: Springer, 2011), accessed on Google Books, https://books.google.com/books?id=Xd6HDE5wUIoC&pg=PA278#v=onepage&q&f=false.

[2] Jessie Chantel, "Smartphone History: The Timeline of a Modern Marvel," Textedly website, February 7, 2024, https://blog.textedly.com/smartphone-history-when-were-smartphones-invented.

[3] "15 Fantastic Firsts on the Internet," Solarwinds website, February 8, 2010, https://www.pingdom.com/blog/15-fantastic-firsts-on-the-internet/.

[4] Michael Phillips, "Reading and Writing Novels by Text," Website magazine, September 22, 2008, https://www.websitemagazine.com/marketing/reading-and-writing-novels-by-text.

[5] "Rocket eBook," MobileRead Wiki, https://wiki.mobileread.com/wiki/Rocket_eBook.

[6] Marie Lebert, "eBooks: 1998—The first ebook readers," Project Gutenberg News, July 16, 2011, http://www.gutenbergnews.org/20110716/ebooks-1998-the-first-ebook-readers/

[7] "Gemstar Acquires NuvoMedia, Inc. and SoftBook Press, Inc." Information Today website, March 2000, https://www.infotoday.com/it/mar00/news12.htm.

[8] "Martin Eberhard," Wikipedia, https://en.wikipedia.org/wiki/Martin_Eberhard.

[9] Brad Stone, *The Everything Store: Jeff Bezos and the Age of Amazon* (New York: Little, Brown, 2013). https://prachititg.com/wp-content/uploads/2014/04/the-everything-store-jeff-bezos-and-the-age-of-amazon.pdf and http://tinyurl.com/mr2hzxhp.

[10] Calvin Reid, "Everybook: The Full-Color E-Book," *Publishers Weekly*, April 26, 1999, https://www.publishersweekly.com/pw/print/19990426/33212-everybook-the-full-color-e-book.html.

6. CONSOLIDATION

[1] Thomas Whiteside, *The Blockbuster Complex: Conglomerates, Show Business, and Book Publishing* (Middletown CT: Wesleyan University Press, 1981).

[2] Ann Crittenden, "Merger Fever in Publishing," *New York Times*, October 23, 1977, https://www.nytimes.com/1977/10/23/archives/merger-fever-in-publishing-merger-fever-in-book-publishing.html.

[3] Rachel Brooke, "The Consolidation of Publishing Houses, Past and Present," Authors Alliance website, December 8, 2021, https://digitalcommons.pace.edu/cgi/viewcontent.cgi?article=1018&context=dyson_mspublishing; also https://www.authorsalliance.org/2021/12/08/the-consolidation-of-publishing-houses-past-and-present/;

Michelle Kratz, *Corporate Influence: How the Media Merger Trend Changed the Book Publishing Industry and the Distribution of Information,* Masters Thesis, Pace University, December 2009, https://digitalcommons.pace.edu/cgi/viewcontent.cgi?article=1018&context=dyson_mspublishing.

[4] Jack Romanos, email to the author dated November 17, 2024; Bob Summer, "PW: Pelican Using Lightning Print to Get Up to Speed," *Publishers Weekly,* August 10, 1998, https://www.publishersweekly.com/pw/print/19980810/25525-pw-pelican-using-lightning-print-to-get-up-to-speed.html; "Successful Pilot Leads Lightning Print Inc. to full launch as Publishers See Benefits from one book at a time service," Ingram Content Group press release, May 26, 1998, https://librarytechnology.org/document/26181.

[5] "Literary Agents' Guild or Association?—AALA/AAR," Literary Agent Undercover website, https://literary-agents.com/association-of-authors-representatives/literary-agents-guild/#:~:text=Sometimes%20mistakenly%20thought%20to%20be,which%20the%20AALA%20does%20not.

[6] Jim Milliot, "Over the Past 25 Years, the Big Publishers Got Bigger—And Fewer," *Publishers Weekly,* April 19, 2022, https://www.publishersweekly.com/pw/by-topic/industry-news/publisher-news/article/89038-over-the-past-25-years-the-big-publishers-got-bigger-and-fewer.html.

[7] Arvyn Cerezo, "What Happens When Publishing Houses Merge?" Book Riot website, November 22, 2023, https://bookriot.com/what-happens-when-publishing-houses-merge/.

[8] "Barnes & Noble, Inc.—Company Profile, Information, Business Description, History, Background Information on Barnes & Noble, Inc." Reference for Business website, https://www.referenceforbusiness.com/history2/8/Barnes-Noble-Inc.html#ixzz7mRBQ4rlG.

[9] Randy Kennedy, "Cash Up Front," *New York Times,* June 5, 2005, https://www.nytimes.com/2005/06/05/books/review/cash-up-front.html?searchResultPosition=1.

[10] Edward Nawotka, "Are Announced First Printings Make-Believe?" Publishing Perspective, August 9, 2011, https://publishingperspectives.com/2011/08/are-announced-first-printings-make-believe/.

[11] Author interview with Power, 2023.

[12] David D. Kirkpatrick, "Book Returns Rise, Signaling a Downturn in the Market," *New York Times,* July 2, 2001, https://www.nytimes.com/2001/07/02/business/book-returns-rise-signaling-a-downturn-in-the-market.html; April Watson, "Book store sales in the United States from 1992 to 2023, Statista, April 9, 2024, https://www.statista.com/statistics/197710/annual-book-store-sales-in-the-us-since-1992/.

[13] Author interview with Power, 2023.

[14] Patricia Holt, "Between the Lines—Independents Win Equal Discounts," *SFGate,* December 10, 1995, https://www.sfgate.com/books/article/BETWEEN-THE-LINES-Independents-Win-Equal-3017551.php.

[15] Janelle Brown, "Barnes & Noble and Borders Sued by Booksellers," *Wired*, March 18, 1998, https://www.wired.com/1998/03/barnes-noble-and-borders-sued-by-booksellers/.

[16] Teresa Méndez, "Why books tours are passé," *Christian Science Monitor,* November 30, 2007, https://www.csmonitor.com/2007/1130/p12s02-bogn.html.

[17] Gerald Howard, "On the Glory Days of the Great American Trade Paperback," *LitHub,* December 15, 2020, https://lithub.com/on-the-glory-days-of-the-great-american-trade-paperback/.

[18] Tony Chiu, "A Record $3.2 Million Is Pledged by Bantam for New Krantz Novel," *New York Times,* September 14, 1979, https://www.nytimes.com/1979/09/14/archives/a-record-32-million-is-pledged-by-bantam-for-new-krantz-novel-in.html.

[19] Jane Austen Doe, "The confessions of a semi-successful author," *Salon,* March 22, 2004, https://www.salon.com/2004/03/22/midlist/.

7. "PUH . . ."

[1] "E-book sales up 68.4% to $113 million in 2008," MobileRead website, April 3, 2009, https://www.mobileread.com/forums/showthread.php?t=44117.

[2] "E-Books Sill Long Way Off from Joining Best-Seller List," *Chicago Tribune,* August 20, 2021, https://www.chicagotribune.com/2000/10/16/e-books-still-long-way-off-from-joining-best-seller-list/; Bill Gates, "Content Is King," January 3, 1996, https://kyrgyzstan.unfpa.org/sites/default/files/pub-pdf/content-is-king.pdf.

[3] Byron Shaw, "Celebrating 25 Years of Xerox Docutech," October 7, 2015, https://byronshaw.wordpress.com/2015/10/07/celebrating-25-years-of-xerox-docutech/.

[4] "Jason Epstein," Wikipedia, https://en.wikipedia.org/wiki/Jason_Epstein; Jason Epstein, *Book Business: Publishing Past, Present and Future* (New York: Norton, 2001); Dinitia Smith, "A Vision for Books That Exults in Happenstance," *New York Times,* January 13, 2001, https://www.nytimes.com/2001/01/13/books/a-vision-for-books-that-exults-in-happenstance.html?scp=12&sq=Robert%20Denning&st=cse.

[5] Judith Rosen, "The Future of the Espresso Book Machine," *Publishers Weekly,* March 4, 2022, https://www.publishersweekly.com/pw/by-topic/industry-news/publisher-news/article/88695-the-future-of-the-espresso-book-machine.html.

[6] Candace Osmond, "What Is Outsource? Meanings and Examples in a Sentence," Grammarist website, https://grammarist.com/usage/outsource/.

[7] "The Evolution of the Conference Call," Ring Central website, January 17, 2023, https://www.ringcentral.com/gb/en/blog/the-evolution-of-the-conference-call/.

8. THE BIG SLEEP

[1] Jim Milliot and Calvin Reid, "Reality Check," *Publishers Weekly,* January 7, 2002, https://www.publishersweekly.com/pw/print/20020107/36156-reality-check.html.

[2] David Kleinbard, "The $1.7 trillion dot.com lesson," *CNN Money,* November 9, 2000, https://money.cnn.com/2000/11/09/technology/overview/.

[3] Chris Gaither and Dawn C. Chmielewski, "Fears of Dot-Com Crash, Version 2.0," *Los Angeles Times,* July 16, 2006, https://www.latimes.com/archives/la-xpm-2006-jul-16-fi-overheat16-story.html.

[4] Associated Press, "Time Warner shuts down e-book division," *Chron,* December 5, 2001, https://www.chron.com/business/technology/article/Time-Warner-shuts-down-e-book-division-2065393.php.

[5] Linds Harrison, "Steven King Reveals *The Plant* Profit," *The Register,* February 7, 2001, https://www.theregister.com/2001/02/07/stephen_king_reveals_the_plant/.

[6] "E-Books Sill Long Way Off from Joining Best-Seller List," *Chicago Tribune.*

[7] Michael Mace, "WhyE-Books Failed in 2000, and What It Means for 2010," *Business Insider,* March 19, 2010, https://www.businessinsider.com/why-ebooks-failed-in-2000-and-what-it-means-for-2010-2010-3.

[8] "Open eBook Forum Updates Industry Specification," Information Today website, October 2001, https://www.infotoday.com/it/oct01/news15.htm.

[9] "Gemstar Acquires NuvoMedia, Inc. and SoftBook Press, Inc." Information Today website.

9. SHIFTING SANDS

[1] Christopher West David, "Want to Do Lunch?" *New York Times,* August 24, 2003, https://www.nytimes.com/2003/08/24/nyregion/in-business-want-to-do-lunch.html.

[2] Interview with Power; "Cader's Media Meal," *Publishing Trends,* September 1, 2000, "https://publishingtrends.com/2000/09/caders-media-meal/.

[3] Publishers Lunch, Friday, November 15, 2024.

10. SLOUCHING TOWARDS KINDLE

[1] "Industry Statistics," International Digital Publishing Forum website, https://idpf.org/about-us/industry-statistics; "The Rise of Ebooks—IDPF Reports Ebook Sales Up 108%," Smashwords blog, January 21, 2009, https://blog.smashwords.com/2009/01/rise-of-ebooks-idpf-reports-ebook-sales.html.

[2] Daniel A. Gross, "The Surprisingly Big Business of Library E-Books," *New Yorker,* September 2, 2021, https://www.newyorker.com/news/annals-of-communications/an-app-called-libby-and-the-surprisingly-big-business-of-library-e-books.

[3] Randall Stross, "Publishers vs. Libraries: An E-Book Tug of War," *New York Times,* December 24, 2011, https://www.nytimes.com/2011/12/25/business/for-libraries-and-publishers-an-e-book-tug-of-war.html.

[4] "Libraries, Patrons, and E-Books. Part I: An introduction to the issues surrounding libraries and e-books," Pew Research Center, June 22, 2012, https://www.pewresearch.org/internet/2012/06/22/part-1-an-introduction-to-the-issues-surrounding-libraries-and-e-books/.

[5] Andrew Albanese, "Macmillan Abandons Library E-Book Embargo," *Publishers Weekly,* March 17, 2020, https://www.publishersweekly.com/pw/by-topic/industry-news/libraries/article/82715-macmillan-abandons-library-e-book-embargo.html.

[6] "Warning: You Are About to Enter the Ebook Zone," *American Libraries,* May 22, 2012, https://americanlibrariesmagazine.org/2012/05/22/warning-you-are-about-to-enter-the-ebook-zone/.

[7] Jon Russell, "Rakuten Buys Ebook and Audiobook Platform Rakuten for $410M," TechChrunch, March 19, 2015, https://techcrunch.com/2015/03/19/rakuten-ovedrive/.

[8] Richard Curtis, "Bullish on E-Books," *Publishers Weekly,* January 7, 2002, https://www.publishersweekly.com/pw/print/20020107/19274-bullish-on-e-books.html.

[9] "Top Selling eBooks of 2004 Announced," International Digital Publishing Forum website, January 12, 2005, https://idpf.org/news/top-selling-ebooks-of-2004-announced.

[10] "PW: Eleven for the Millennium," *Publishers Weekly,* January 3, 2000, https://www.publishersweekly.com/pw/print/20000103/16557-pw-eleven-for-the-millennium.html.

[11] *Inside,* March 20, 2001.

11. THE ROAD TO E-DAY

[1] John Hollar, interviewer, *Oral History of Greg Zehr,* Computer History Museum, May 9, 2014.

[2] Zach Pontz, "A year later, Amazon's Kindle finds a niche," CNN, N.D., https://www.cnn.com/2008/TECH/12/03/kindle.electronic.reader/; Michael Kozlowski, "Documentary—The entire history of Sony e-readers and e-notes," Good e-Reader website, August 17, 2022, https://goodereader.com/blog/electronic-readers/good-e-reader-documentary-the-entire-history-of-sony-e-readers-and-e-notes.

[3] John Hollar, interviewer, *Oral History of Greg Zehr.*

[4] Neal Karlinsky, "The inside story of how the Kindle was born," Amazon News website, November 15, 2017, https://www.aboutamazon.com/news/devices/the-inside-story-of-how-the-kindle-was-born.

[5] Walter Isaacson, *Steve Jobs* (New York: Simon & Schuster, 2013), page 576.

[6] Erick Schonfeld, "We Know How Many Kindles Amazon Has Sold: 240,000," TechCrunch website, August 1, 2008, https://techcrunch.com/2008/08/01/we-know-how-many-kindles-amazon-has-sold-240000/.

[7] David Carnoy, "Report: Amazon likely to sell 8 million Kindles in 2010," December 21, 2010, https://www.cnet.com/culture/report-amazon-likely-to-sell-8-million-kindles-in-2010/.

[8] Amy Watson, "Number of Barnes & Noble stores from fiscal year 2005 to fiscal year 2019," Statista, July 17, 2020, https://www.statista.com/statistics/199012/number-of-barnes-noble-stores-by-type-and-year-since-2005/.

[9] Definition of "nooky," Merriam-Webster Dictionary, https://www.merriam-webster.com/dictionary/nooky.

12. GAME ON

[1] Nilay Patel, "Kindle sells out in 5.5 hours," Engadget website, November 21, 2007, https://www.engadget.com/2007-11-21-kindle-sells-out-in-two-days.html.

[2] John Sargent, *Turning Pages* (New York: Arcade Publishing, 2023), page 79.

[3] Nilay Patel, "Kindle sells out in 5.5 hours."

[4] Alex Wilhelm, "How Many Kindles Have Been Sold?" TNW website, July 29, 2010, https://thenextweb.com/news/how-many-kindles-have-been-sold.

[5] Jason Del Ray, "How Big Is Amazon's Kindle Business? Morgan Stanley Takes a Crack at It," All Things Digital website, August 12, 2003, https://allthingsd.com/20130812/amazon-to-sell-4-5-billion-worth-of-kindles-this-year-morgan-stanley-says/?mod=obinsite.

[6] Michael Kozlowski, "The History of the Barnes and Noble Nook and eB ook Ecosystem," GoodEReader website, October 12, 2012, https://goodereader.com/blog/electronic-readers/the-history-of-the-barnes-and-noble-nook-and-ebook-ecosystem.

[7] Dean van Leeuwen, "The Quest for a Great Computer in a Book," Tomorrow Today website, April 21, 2017, https://tomorrowtodayglobal.com/2017/04/21/quest-great-computer-book/.

[8] Dawn Kawamoto, "Newsmaker: Riding the next technology wave," CNET News, October 2, 2003, https://web.archive.org/web/20120205201219/http://news.cnet.com/2008-7351-5085423.html.

[9] "Reading ebooks on an iPod," Digital Bits Technology Column website, http://www.andybrain.com/archive/ebooks-on-ipod.htm.

[10] Luke Dormehl, "Today in Apple history: Apple sells its 100 millionth iPod," Cult of Mac website, April 9, 2025, https://www.cultofmac.com/539643/100-million-ipods-sold/.

[11] "Annual Sales of Apple's iPhone (2007-2021)," Global data website, https://www.globaldata.com/data-insights/technology--media-and-telecom/annual-sales-of-apples-iphone/.

[12] Roger Fingas, "A brief history of the iPad, Apple's once and future tablet," Apple Insider website, April 3, 2018, https://appleinsider.com/articles/18/04/03/a-brief-history-of-the-ipad-apples-once-and-future-tablet.

[13] "Global Apple iPad sales from 3rd fiscal quarter of 2010 to 4th fiscal quarter of 2018 (in million units)," Statista, https://www.statista.com/statistics/269915/global-apple-ipad-sales-since-q3-2010/.

[14] "iPad," Wikipedia, https://en.wikipedia.org/wiki/IPad.

[15] Daniel Nations, "How Many iPads Have Been Sold?" Lifewire website, September 11, 2024, https://www.lifewire.com/how-many-ipads-sold-1994296.

[16] "Apple Books: What's new in iOS12," iMore website, September 22, 2018, https://www.imore.com/apple-books-whats-new-ios-12.

[17] Calvin Reid, "Apple Says iBooks Store Attracts 1 Million Users a Week," *Publishers Weekly,* January 15, 2015 https://www.publishersweekly.com/pw/by-topic/digital/retailing/article/65295-apple-says-ibooks-store-attracts-1-million-users-a-week.html.

[18] Steve Zeitchik, "Byron Preiss: 'He Saw Books Where Other People Didn't,'" *Publishers Weekly,* July 15, 2005, https://www.publishersweekly.com/pw/print/20050718/40390-byron-preiss-he-saw-books-where-other-people-didn-t.html.

[19] Steve Zeitchik, "Byron Preiss: 'He Saw Books Where Other People Didn't.'"

[20] Augie De Blieck, Jr., 'Blacksad and ibooks: The Incredible True Story of the Publisher That Brought Blacksad to America, Died, and Sued Apple (In That Order)," Pipeline Comics website, August 7, 2021, https://www.pipelinecomics.com/blacksad-ibooks-byron-preiss/#google_vignette.

[21] Glynnis MacNicol, "Former HarperCollins CEO Jane Friedman Secures $8 Million in Funding for E-Book Business," *Business Insider,* May 23, 2011, https://www.businessinsider.com/jane-friedman-harper-ebook-open-road-2011-5.

[22] "Harlan Ellison—Pay the Writer," video, YouTube, https://www.youtube.com/watch?v=PuLr9HG2ASs.

[23] "The Harlan Ellison Collection on E-Reads," video, YouTube, https://www.youtube.com/watch?v=IbOBR3MQ_F4.

[24] Emails among Danny Baror, agent; Philip Patrick, Amazon.com; Richard Curtis; and Larry Kirshbaum, Amazon.com, dated July 25, 2011 to August 11, 2011.

[25] Josh Sanburn, "In Latest Moves, Barnes & Noble is Betting it Can Compete With Amazon," *Time,* February 6, 2012, https://business.time.com/2012/02/06/in-latest-moves-barnes-noble-is-betting-it-can-compete-with-amazon/.

[26] Jeffrey Dastin, "Amazon to shut its bookstores and other shops as its grocery chain expands," Reuters, March 2, 2022, https://www.reuters.com/business/retail-consumer/exclusive-amazon-close-all-its-physical-bookstores-4-star-shops-2022-03-02/.

[27] Stephen Silver, "The revolution Steven Jobs resisted: Apple's App Store marks ten years of third-party innovation," Apple Insider, July 10, 2018,

https://appleinsider.com/articles/18/07/10/the-revolution-steve-jobs-resisted-apples-app-store-marks-10-years-of-third-party-innovation.

[28] Shehraj Singh, "Kindle Users and Usage Statistics,"eReader blog, July 18, 2023, https://ereader.blog/kindle-statistics/.

[29] Amy Watson, "Estimated number of e-books sold in the United States from 2010 to 2020," Statista, September 8, 2023, https://www.statista.com/statistics/426799/e-book-unit-sales-usa/.

[30] Edward Nawotka, "Is There a Right Way to Feel About Amazon?" *Publishing Perspectives,* April 5, 2012, https://publishingperspectives.com/2012/04/is-there-a-right-way-to-feel-about-amazon/.

[31] Peter Osnos, "What Went Wrong at Borders," *The Atlantic,* January 11, 2011, https://www.theatlantic.com/business/archive/2011/01/what-went-wrong-at-borders/69310/; Eyder Peralta, "Headed For Liquidation, Borders Will Close Its Doors," NPR, July 18, 2011, https://www.npr.org/sections/thetwo-way/2011/07/18/138491830/headed-for-liquidation-borders-will-close-its-doors.

[32] Yuki Noguchi, "Why Borders Failed While Barnes & Noble Survived," *All Things Considered,* July 19, 2011, https://www.npr.org/2011/07/19/138514209/why-borders-failed-while-barnes-and-noble-survived.

[33] Rhonda Abrams, "Strategies: 10 ways you can combat showrooming," *USA Today,* April 12, 2013, https://www.usatoday.com/story/money/columnist/abrams/2013/04/12/small-business-online-frustrations/2076169/.

[34] Eyder Peralta, "Headed For Liquidation, Borders Will Close Its Doors."

[35] Nemanja Curcic, "How much does Amazon make a day?" Finbold website, January 6, 2025, https://finbold.com/guide/how-much-does-amazon-make-a-day/; Dean Talbot, "Book Publishing Companies Statistics," WordsRated website, January 27, 2023, https://wordsrated.com/book-publishing-companies-statistics/.

13. WHAT WERE THEY THINKING?

[1] Adam Rowe, "U.S. Publishers Are Still Losing $300 Million Annually to Ebook Piracy," *Forbes,* July 28, 2019, https://www.forbes.com/sites/adam-rowe1/2019/07/28/us-publishers-are-still-losing-300-million-annually-to-ebook-piracy/?sh=50383bd3319e.

[2] Kassia Krozser, "Delaying Ebook Releases: A Publisher Weighs In," Booksquare website, July 15, 2009, https://booksquare.com/delaying-ebook-releases-a-publisher-weighs-in/.

[3] Hannah Johnson, "One-star Ratings on Amazon for Book Without Kindle Version," *Publishing Perspectives,* January 18, 2010, https://publishingperspectives.com/2010/01/one-star-ratings-on-amazon-for-book-without-kindle-version/.

[4] Steven W. Beattie, "Unhappy Kindle users unleash a flood of one-star reviews for title with no e-book edition," *Quill & Quire,* January 18, 2010,

https://quillandquire.com/book-news/2010/01/18/unhappy-kindle-users-unleash-a-flood-of-one-star-reviews-for-title-with-no-e-book-editon/.

[5] Adrian Covert, "A decade of iTunes singles killed the music industry, *CNN Business,* April 25, 2013, https://money.cnn.com/2013/04/25/technology/itunes-music-decline/; Glenn Rifkin, "How Tower Records Spun Out of Business," KornFerry website, April 17, 2020, https://www.kornferry.com/insights/briefings-magazine/issue-43/how-tower-records-spun-out-of-business.

[6] Dennis Abrams, "Amazon Explains Why It Thinks $9.99 Is the Right Price for Ebooks," *Publishing Perspectives,* July 31, 2014, https://publishingperspectives.com/2014/07/amazon-explains-why-it-thinks-9-99-is-the-right-price-for-ebooks/.

[7] The legal quotations in this chapter are taken from the transcript of *United States v. Apple Inc.,* 952 F. Supp. 2d 638 (S.D.N.Y. 2013), https://www.leagle.com/decision/infdco20130711c57.

[8] *United States v. Apple Inc.*

[9] Brad Stone and Motoko Rich, "Amazon Removes Macmillan Books," *New York Times,* January 30, 2010, https://www.nytimes.com/2010/01/30/technology/30amazon.html?auth=login-email&login=email.

[10] Jane Little, "DOJ Lawsuit Update: Where Windowing Becomes Important," Teleread website, May 15, 2012, https://teleread.com/doj-lawsuit-update-where-windowing-becomes-important-by-jane-litte/index.html.

[11] "Amazon stock price in 2010," Statmuse website, https://www.statmuse.com/money/ask/amazon+stock+price+in+2010.

[12] Charlie Osborne, "Apple settles ebook antitrust case, set to pay millions in damages," ZDNet website, June 17, 2014, https://www.zdnet.com/article/apple-settles-ebook-antitrust-case-set-to-pay-millions-in-damages/.

[13] Charlie Osborne, "Apple settles ebook antitrust case, set to pay millions in damages."

[14] John Hollar, interviewer, *Oral History of Greg Zehr.*

[15] Jim Milliot, "Book Business Applauds Government Lawsuit Against Amazon," *Publishers Weekly,* September 27, 2023, https://www.publishersweekly.com/pw/by-topic/industry-news/bookselling/article/93271-book-business-applauds-government-lawsuit-against-amazon.html.

14. DISPLACED PERSONS

[1] Colin Meloy, "Midlist Author," sung by The Decembrists, https://www.google.com/search?client=firefox-b-1-d&q=midlist+author+lyrics.

[2] "How much does a book editor make in New York, NY?" Salary.com website, May 1, 2025, https://www.salary.com/research/salary/recruiting/book-editor-salary/new-york-ny.

[3] Katherine Rosman, "The Death of the Slush Pile," *Wall Street Journal,* January 22, 2010, https://www.wsj.com/articles/SB10001424052748703414504575001271351446274.

[4] Kate McKean, "A Year of Queries," Agents and Books, Substack, October 1, 2024, https://katemckean.substack.com/p/a-year-of-queries.

[5] Michael Coffey, "The Myth Of the Gatekeepers," *Publishers Weekly,* May 23, 2014, https://www.publishersweekly.com/pw/by-topic/columns-and-blogs/soapbox/article/62413-the-myth-of-the-gatekeepers.html.

[6] "BookSurge, inventory-free global publishing platform," Failory website, https://www.failory.com/amazon/booksurge.

[7] Jim Milliot, "BookSpace, CreateSpace Merge," *Publishers Weekly,* December 1, 2009, https://www.publishersweekly.com/pw/by-topic/industry-news/publisher-news/article/24989-booksurge-createspace-merge.html

[8] Amy Watson, "Number of books self-published via CreateSpace in the United States from 2010 to 2018," Statista, August 13, 2024, https://www.statista.com/statistics/605000/createspace-number-books-published/.

[9] Judith Rosen & Jim Milliot, "With Preint Book Sales Stabilized, Return Rate Lowers," *Publishers Weekly,* May 6, 2016, https://www.publishersweekly.com/pw/by-topic/industry-news/bookselling/article/70298-with-print-book-sales-stabilized-return-rate-lowers.html.

[10] Alison Flood, "Publishing suffers its first casualties of the recession," *The Guardian,* December 5, 2008, https://www.theguardian.com/books/booksblog/2008/dec/05/publishing-recession.

[11] Novella Carpenter, "Book publishers, R.I.P.? In this bad economy, it's tougher than ever to sell books," *SFGate,* February 24, 2009, https://www.sfgate.com/business/article/Book-publishers-R-I-P-In-this-bad-economy-2481695.php.

[12] Nicholas Rizzo, "Self-published Books & Authors Sales Statistics [2023]," WordsRated website, January 30, 2023, https://wordsrated.com/self-published-book-sales-statistics/.

[13] J. A. Konrath, "Do Legacy Publishers Treat Authors Badly?" J. A. Konrath blog, February 20, 2012, http://jakonrath.blogspot.com/2012/02/do-legacy-publishers-treat-authors.html.

[14] J. A. Konrath, "Unconscionability," J. A. Konrath blog, May 23, 2012, https://jakonrath.blogspot.com/2012/05/unconscionability.html.

[15] Carolyn Kellogg, "Independent author John Locke joins Amazon's million-Kindle-seller club, but at what cost?" *Los Angeles Times,* June 21, 2011, https://www.latimes.com/archives/blogs/jacket-copy/story/2011-06-21/independent-author-john-locke-joins-amazons-million-kindle-seller-club-but-at-what-cost.

[16] David Ferrell, "From Insurance Salesman to Bestselling Ebook Author," Second Act website, September 12, 2011,

https://web.archive.org/web/20130608081257/http://www.sec-ondact.com/2011/09/insurance-salesman-becomes-bestselling-ebook-author/.

[17] Hugh Howey, "How WOOL Got A Unique Publishing Deal," *HuffPost*, March 11, 2013, https://www.huffpost.com/entry/how-wool-got-a-unique-pub_b_2852547.

[18] "Joe Konrath signs with Legacy Publishing House: Hypocrisy or shrewd business opportunity?" Gary Dobbs / Jack Martin blog, https://tainted-archive.blog-spot.com/2018/03/joe-konrath-signs-with-legacy.html?m=1.

[19] Lauren Oliver, "The Relaunch of Amanda Hocking," *New York Times*, January 13, 2012, https://www.nytimes.com/2012/01/15/books/review/switched-by-amanda-hocking-book-review.html?_r=1&scp=1&sq=amanda%20hocking&st=cse; interview with Axelrod.

[20] Jason Ward, "Is Self-Publishing the New Slush Pile or The End of Publishing?" Medium: The Writing Cooperative, May 30, 2020, https://writingcooperative.com/is-self-publishing-the-new-slush-pile-or-the-end-of-publishing-9530d54d0072.

[21] "Influencer," Sprout Social website, https://sproutsocial.com/glossary/influ-encer/.

[22] Mark Coker, "Hugh Howey and the Indie Author Revolt," *Publishers Weekly*, February 20, 2014, https://www.publishersweekly.com/pw/by-topic/digital/content-and-e-books/article/61116-hugh-howey-and-the-indie-author-revolt.html.

[23] Ben Paynter, "Remember Zagat? The iconic burgundy guidebook that helped shape the modern consumer industry is back," *Fast Company*, October 1, 2010, https://www.fastcompany.com/90227760/remember-zagat-the-iconic-burgundy-guide-book-that-helped-shape-the-modern-consumer-era-is-back.

[24] Stuart A. Thompson, "Fake Reviews Are Rampant Online. Can a Crackdown End Them?" *New York Times*, November 13, 2023, https://www.ny-times.com/2023/11/13/technology/fake-reviews-crackdown.html.

[25] "Submit a Review," Amazon Community website, https://www.ama-zon.com/gp/help/customer/display.html?nodeId=GL4WJF8BGV8VL6B8; "Community Guidelines," Amazon Community website, https://www.amazon.com/gp/help/cus-tomer/display.html?nodeId=201929730.

[26] Alison Flood, "Amazon purchase of Goodreads stuns book industry," *The Guard-ian*, April 2, 2013, https://www.theguardian.com/books/2013/apr/02/amazon-pur-chase-goodreads-stuns-book-industry.

[27] Alison Flood, "Amazon purchase of Goodreads stuns book industry."

[28] "Was Goodreads better before it was bought by Amazon?" Goodreads discussion page, https://www.goodreads.com/topic/show/19883316-was-goodreads-better-be-fore-it-was-bought-by-amazon.

[29] Cybil, "Goodreads Members' Top 72 Hit Books of the Year (So Far)," Goodreads website, June 6, 2022, https://www.goodreads.com/blog/show/2302-goodreads-mem-bers-top-72-hit-books-of-the-year-so-far /.

[30] Publishers Lunch 12/12/20, https://lunch.publishersmarket-place.com/2023/12/del-rey-removes-cait-corrains-book-from-2024-schedule-follow-ing-goodreads-controversy/.

[31] Maris Kreizman, "Let's Rescue Book Lovers From This Online Hellscape," *New York Times*, December 24, 2023, https://www.nytimes.com/2023/12/24/opin-ion/goodreads-books-reviews.html?searchResultPosition=1.

[32] "PW Select FAQs," BookLife website, May 9, 2025, https://booklife.com/about-us/pw-select-faqs.html.

[33] Alan Scherstuhl, "Just Do It (Yourself): A History of Self-Publishing," *Publishers Weekly*, April 19, 2022, https://www.publishersweekly.com/pw/by-topic/industry-news/publisher-news/article/88987-just-do-it-yourself-a-history-of-self-publish-ing.html

15. THE DARK SIDE

[1] "Global Napster Usage Plummets, But New File-Sharing Alternatives Gaining Ground, Reports Jupiter Media Metrix," ComScore Press Center Home, July 20, 2001, https://web.archive.org/web/20080413104420/-http://www.comscore.com/press/re-lease.asp?id=249.

[2] Commission on the Theft of American Intellectual Property, *Update to the IP Commission Report*, February 2017, https://www.nbr.org/wp-content/up-loads/pdfs/publications/IP_Commission_Report_Update.pdf.

[3] Jenna Zaza, "The FBI cracks down on pirated e-book libraries," *Stony Book Press*, February 17, 2023, https://sbpress.com/2023/02/the-fbi-cracks-down-on-pirated-e-book-libraries/.

[4] "Attributor Corporation," CBInsights website, https://www.cbin-sights.com/company/attributor-corporation.

[5] Katy Guest, "'I can get any novel I want in 30 seconds': can book piracy be stopped?" *The Guardian*, March 6, 2019, https://www.theguard-ian.com/books/2019/mar/06/i-can-get-any-novel-i-want-in-30-seconds-can-book-pi-racy-be-stopped; Sovan Mandal, "How an e-book is pirated, its implications for the stake-holders, and the extent of the problem," *Good e-Reader*, April 13, 2023, https://good-ereader.com/blog/e-book-news/how-an-e-book-is-pirated-its-implications-for-the-stakeholders-and-the-extent-of-the-problem.

[6] Allan Ryan, "Hack Your Amazon Kindle 4th Generation E-reader," YouTube, https://www.youtube.com/watch?v=CbtVY8vmqmM; Reuters, "Hackers claim victory in cracking Amazon Kindle DRM," Decemver 24, 2009, https://www.reuters.com/arti-cle/urnidgns002570f3005978d8002576950064f79a/hackers-claim-victory-in-cracking-amazon-kindle-drm-idUS268769666420091224.

[7] "The Dangerous World of Counterfeit and Pirated Software," Microsoft white paper, March 2013, https://news.microsoft.com/download/presskits/antipiracy/docs/IDC030513.pdf.

[8] Sam Law, "Metallica vs. Napster: The lawsuit that redefined how we listen to music," *Kerrang* magazine, April 13, 2021, https://www.kerrang.com/metallica-vs-napster-the-lawsuit-that-redefined-how-we-listen-to-music.

[9] R. Polk Wagner, "Information Wants To Be Free: Intellectual Property and the Mythologies of Control," *Columbia Law Review*, May 9, 2003, https://web.archive.org/web/20101226054908/http://www.law.upenn.edu/fac/pwagner/wagner.control.pdf.

[10] "Authors Guild Issues Report Exploring the Factors Leading to the Decline of the Writing Profession," Authors Guild website, February 19, 2020, https://authorsguild.org/news/authors-guild-issues-report-exploring-the-factors-leading-to-the-decline-of-the-writing-profession/; Niall McCarthy, "U.S. Authors Have Suffered a Drastic Decline in Earnings," *Forbes*, January 9, 2019, https://www.forbes.com/sites/niallmccarthy/2019/01/09/u-s-authors-have-suffered-a-drastic-decline-in-earnings-infographic/.

[11] Institute for Policy Innovation, "The True Cost of Sound Recording Piracy to the U.S. Economy," Recording Industry Association of America website, https://www.riaa.com/reports/the-true-cost-of-sound-recording-piracy-to-the-u-s-economy/.

[12] Jack Lynch, "The Perfectly Acceptable Practice of Literary Theft: Plagiarism, Copyright, and the Eighteenth Century," *Writing-World.com*, https://www.writing-world.com/rights/lynch.shtml.

[13] Dian Schaffhauser, "Report: Students Plagiarized More When Instruction Moved Online," *Technological Horizons in Education*, April 9, 2021, https://thejournal.com/articles/2021/04/09/report-students-plagiarized-more-when-instruction-moved-online.aspx.

[14] "AI Tools and Resources," University of South Florida Libraries website, https://guides.lib.usf.edu/c.php?g=1315087&p=9678778. Reflecting how quickly our adaptation to AI is evolving, the wording on this site had changed within a few months of my drafting of this chapter, having been updated May 6, 2025.

[15] "Statement on AI training," https://www.aitrainingstatement.org.

[16] Adam W. Sikich, "Fair or Foul? The Unanswered Fair Use Implications of the Google Library Project," *Landslide*, Volume 2, Number 1, September/October 2009, http://dunnerlaw.com/wp-content/uploads/2016/12/Google-article.pdf.

[17] "Google hits back at book critics," BBC News, October 9, 2009, http://news.bbc.co.uk/1/hi/technology/8298674.stm.

[18] "Authors Guild v. Google, Part II: Fair Use Proceedings," Electronic Frontier Foundation website, https://www.eff.org/cases/authors-guild-v-google-part-ii-fair-use-proceedings.

[19] Joseph Ax, "Google book-scanning project legal, says U.S. appeals court," Reuters, October 16, 2015, https://www.reuters.com/article/us-google-books-idUSKCN0SA1S020151016/; Robinson Meyer, "After 10 Years, Google Books Is Legal," *The Atlantic*, October 20, 2015, https://www.theatlantic.com/technology/archive/2015/10/fair-use-transformative-leval-google-books/411058/.

16. INDELIBLE INK

[1] Amy Watson, "U.S. book market—statistics & facts," Statista, May 16, 2024, https://www.statista.com/topics/1177/book-market/#topicOverview.

[2] Virginia Heffernan, "Watch Me, Read Me," *New York Times Magazine,* January 14, 2011, https://www.nytimes.com/2011/01/16/magazine/16FOB-medium-t.html.

[3] Tiffany G. Munzer, et. al., "Differences in Parent-Toddler Interactions With Electronic Versus Print Books," *Pediatrics* (2019) 143 (4): e20182012, https://publications.aap.org/pediatrics/article/143/4/e20182012/76785/Differences-in-Parent-Toddler-Interactions-With?autologincheck=redirected.

[4] Jim Milliot, "BEA 2013: The E-Book Boom Years," *Publishers Weekly*, May 29, 2013, https://www.publishersweekly.com/pw/by-topic/industry-news/bea/article/57390-bea-2013-the-e-book-boom-years.html.

[5] "Children's Book Publishing in the US—Market Size (2005-2030)," IBISWorld website, October 2024, https://www.ibisworld.com/industry-statistics/market-size/childrens-book-publishing-united-states/.

[6] Alexandra Alter, "The Hottest Trend in Publishing: Books You Can Judge by Their Cover," *New York Times,* December 27, 2024, https://www.nytimes.com/2024/12/27/books/deluxe-book-editions-decorated-edges.html?searchResultPosition=1.

[7] Jin Milliot, "Readerlink Will Stop Distributing Mass Market Paperbacks at the End of 2025," *Publishers Weekly,* February 4, 2025, https://www.publishersweekly.com/pw/by-topic/industry-news/industry-deals/article/97161-readerlink-will-stop-distributing-mass-market-paperbacks-at-the-end-of-2025.html.

[8] Jim Milliot, "Book Sales Continue to Slow Down in First Half of 2023," *Publishers Weekly,* July 7, 2023, https://www.publishersweekly.com/pw/by-topic/industry-news/bookselling/article/92735-book-sales-continue-to-slow-down-in-first-half-of-2023.html.

[9] Jim Milliot, "Where Are Mass Market Paperbacks Headed?" *Publishers Weekly,* August 5, 2022, https://www.publishersweekly.com/pw/by-topic/industry-news/bookselling/article/90039-mass-market-paperback-sales-whither.html.

[10] Adam Hayes, "Long Tail: Definition as a Business Strategy and How It Works," *Investopedia,* January 26, 2025, https://www.investopedia.com/terms/l/long-tail.asp.

[11] Jim Milliot, "How Ingram Content Group Became a $2 Billion Business," *Publishers Weekly,* April 16, 2021, https://www.publishersweekly.com/pw/by-topic/industry-

news/publisher-news/article/86111-how-ingram-content-group-became-a-2-billion-company.html.

[12] Jim Milliot, "Pietsch Touts New Marketing Efforts in Letter to Authors," *Publishers Weekly*, December 26, 2023, https://www.publishersweekly.com/pw/by-topic/industry-news/publisher-news/article/94007-pietsch-touts-new-marketing-efforts-in-letter-to-authors.html

[13] Thad McIlroy and Jim Milliot, "Over 30 Years, 40% of Publishing Jobs Disappeared. What Happened?" *Publishers Weekly*, September 20, 2024, https://www.publishersweekly.com/pw/by-topic/industry-news/publisher-news/article/95996-over-30-years-40-of-publishing-jobs-disappeared-what-happened.html.

[14] Thad McIlroy and Jim Milliot, "Over 30 Years, 40% of Publishing Jobs Disappeared. What Happened?"

[15] Julie *Bosman, "Penguin Acquires Self-Publishing Company," New York Times,* July 19, 2012, https://archive.nytimes.com/mediadecoder.blogs.nytimes.com/2012/07/19/penguin-acquires-self-publishing-company/.

[16] Calvin *Reid, "Pearson Acquires Self-Publishing Vendor Author Solutions for $116 Million,"* July 19, 2012, https://www.publishersweekly.com/pw/by-topic/industry-news/publisher-news/article/53077-pearson-acquires-self-publishing-vendor-author-solutions-for-116-million.html,

[17] "Simon & Schuster Creates Self-Publishing Unit, Archway Publishing," *Publishers Weekly*, November 27, 2012, https://www.publishersweekly.com/pw/by-topic/industry-news/industry-deals/article/54883-simon-schuster-creates-self-publishing-unit-archway-publishing.html.

[18] Jim Milliot, "Author Solutions Sold to Private Equity Firm," *Publishers Weekly*, January 5, 2016, https://www.publishersweekly.com/pw/by-topic/industry-news/publisher-news/article/69062-author-solutions-sold-to-private-equity-firm.html.

[19] "Book Expo America 2012: All Our Coverage," *Publishers Weekly*, April 27, 2012, https://www.publishersweekly.com/pw/by-topic/industry-news/bea/article/51746-book-expo-america-2012-all-our-coverage.html.

[20] Jim Milliot, "BookExpo America 2010: 21,919 'Verified' Attendees at This Year's Event," *Publishers Weekly*, June 2, 2010, https://www.publishersweekly.com/pw/by-topic/industry-news/bea/article/43395-bookexpo-america-2010-21-919-verified-attendees-at-this-year-s-event.html.

[21] Porter Anderson, "BookExpo Announces a Shorter Trade Show for 2020 in New York City," *Publishing Perspectives*, September 9, 2019, https://publishingperspectives.com/2019/09/bookexpo-announces-changes-for-new-york-city-2020-trade-show/.

17. FOR SALE

[1] Jim Milliot, "BEA 2013: The E-Book Boom Years."

[2] Amy Watson, "NOOK sales in the fiscal years 2010 to 2019," Statista, July 17, 2020, https://www.statista.com/statistics/237684/barnes-and-noble-nook-sales/.

[3] Jim Milliot, "Investment Group Led by David Steinberger Buys Open Road," *Publishers Weekly*, December 2, 2021, https://www.publishersweekly.com/pw/by-topic/industry-news/industry-deals/article/88033-investment-group-led-by-david-steinberger-buys-open-road.html.

18. RESCUED

[1] Lt. Col. Dave Grossman, "Hope on the Battlefield," *Greater Good* magazine, June 1, 2007, https://greatergood.berkeley.edu/article/item/hope_on_the_battlefield.

[2] David Martin, "Random Facts About Soldiers Trained to Kill," *Eastern Shore Post*, August 5, 2022, https://easternshorepost.com/2022/08/05/random-facts-about-soldiers-trained-to-kill-part-ii/.

[3] Adapted from tribute to Greg Bear in *Locus* magazine, January 2023, https://locusmag.com/2023/01/issue-744-table-of-contents-january-2023/.

[4] Margalit Fox, "Richard S. Prather, Author of Naked Mysteries, Dies at 85," *New York Times*, March 17, 2007, https://www.nytimes.com/2007/03/17/books/17prather.html.

19. THE END

[1] Originally published in the 2018 Association of Authors' Representatives Newsletter under the title "Goodbye to All That."

20. BACK FROM THE FUTURE

[1] Jim Milliot, "Some Parting Words for the Book Biz from Jim Milliot," *Publishers Weekly*, December 15, 2023, https://www.publishersweekly.com/pw/by-topic/columns-and-blogs/soapbox/article/93954-some-parting-words-for-the-book-biz-from-jim-milliot.html.

[2] Richard Curtis, "Let's Make Some Noise," *Publishers Weekly*, June 29, 2013, https://www.publishersweekly.com/pw/by-topic/columns-and-blogs/soapbox/article/58035-let-s-make-some-noise.html.

[3] Nate Hoffelder, "AAP: eBook Sales Up 41% in 2012 as Growth Slows Down"; Gary Price, "Association of American Publishers Reports (AAP) Overall Publishing Industry Down 6.4% for Calendar Year 2022, and 10.0% for Month of December," Info Docket web site, February 10, 2023, https://www.infodocket.com/2023/02/10/association-of-american-publishers-reports-aap-overall-publishing-industry-down-6-4-for-calendar-year-2022-and-10-0-for-month-of-december/.

[4] Nicholas Rizzo, "Self-published Books & Authors Sales Statistics [2023]," Words Rated website, January 30, 2023, https://wordsrated.com/self-published-book-sales-statistics/.

[5] Derek Haines, "E-Reader Device Sales Including Kindle Are in Decline," August 19, 2024, Just Publishing Advice website, https://justpublishingadvice.com/the-e-reader-device-is-dying-a-rapid-death/.

[6] http://myereader.net/kindle-ereader-sales.html.

[7] Steven Knight, "How Many iPads Have Been Sold? 2023 Statistics," Sellcell website, June 21, 2023, https://www.sellcell.com/blog/how-many-ipads-have-been-sold-2023-statistics/.

[8] Amy Watson, "U.S. book market—statistics & facts," Statista, May 16, 2024, https://www.statista.com/topics/1177/book-market/#topicOverview; "E-reader Market Analysis," Technavio website, January 2024, https://www.technavio.com/report/e-reader-market-industry-analysis.

[9] Audiobook Market Size & Trends 2018-2024, Grand View Research website, https://www.grandviewresearch.com/industry-analysis/audiobooks-market.

[10] "Podcast Statistics You Need to Know," Backlinko website, March 25, 2025, https://backlinko.com/podcast-stats.

[11] "Authors Guild Survey Shows Drastic 42 Percent Decline in Authors Earnings in Last Decade," Authors Guild website, January 5, 2019, https://authorsguild.org/news/authors-guild-survey-shows-drastic-42-percent-decline-in-authors-earnings-in-last-decade/.

[12] "Key Takeaways from the Authors Guild's 2023 Author Income Survey," Authors Guild website, September 27, 2023, https://authorsguild.org/news/key-takeaways-from-2023-author-income-survey/.

[13] "7 Things to Know Before Creating an Audiobook," Author Learning Center website, https://www.authorlearningcenter.com/publishing/formats/w/audiobooks/6327/7-things-to-know-before-creating-an-audiobook-article.

[14] Kim Scott, "Remember What Spotify Did to the Music Industry? Book Are Next," *New York Times,* December 13, 2023, https://www.nytimes.com/2023/12/13/opinion/audiobooks-spotify-streaming-algorithm.html?smid=nytcore-android-share.

[15] Grace Harmon, "Microsoft and Harper Collins sign AI licensing deal, but author opt-in still required," *EMarketer*, November 25, 2024, https://www.emarketer.com/content/microsoft-harpercollins-sign-ai-licensing-deal--author-opt-in-still-required.

[16] Lauren Harvey, "Copyrighted books are fair use for AI training, federal judge rules in Anthropic case, *Los Angeles Times,* June 25, 2025, https://www.latimes.com/entertainment-arts/story/2025-06-25/anthropic-copyrighted-books-ai-fair-use.

[17] Kate Knibbs, "Thomson Reuters Wins First Major AI Copyright Case in the US," *Wired,* February 11, 2025, https://www.wired.com/story/thomson-reuters-ai-copyright-lawsuit/.

SOURCE NOTES

[18] "AI investment forecast to approach $200 billion globally by 2025," Goldman Sachs, August 1, 2023, https://www.goldmansachs.com/insights/articles/ai-investment-forecast-to-approach-200-billion-globally-by-2025.

[19] Jim Milliot, "Authors Guild Reinforces Its Position on AI Licensing," *Publishers Weekly*, December 16, 2024, https://www.publishersweekly.com/pw/by-topic/industry-news/licensing/article/96745-authors-guild-reinforces-its-position-on-ai-licensing.html.

[20] "Open Letter to Generative AI Leaders," Authors Guild website, https://action-network.org/petitions/authors-guild-open-letter-to-generative-ai-leaders.

INDEX

INDEX

INDEX

ABOUT THE AUTHOR

R ICHARD CURTIS is a leading New York literary agent, publishing authority, e-book pioneer, and authors' advocate. While running his eponymous
literary agency, he wrote numerous columns, blogs and articles for *Publishers
Weekly* and other writers' publications, leading to four information-packed
books about writing, agenting and the book business. He was the first president
of the Independent Literary Agents Association and subsequently president of
the Association of Authors' Representatives.

Curtis's fascination with emerging media and technology led to his founding one of the first commercial e-book publishers—seven years prior to the introduction of the Kindle and the advent of the E-Book Revolution. He developed
e-book business and royalty accounting models that are still in use today. His
popular blog, *Publishing in The Twenty-First Century*, describing the wonders and
challenges of the digital paradigm, was followed by both professionals and lay
audiences.

Curtis is also author of dozens of works of fiction and nonfiction. His sa-
tirical end-of-year verses for *Publishers Weekly*, published over a span of 45
years, made him the unofficial poet laureate of the book industry. His plays have
been performed in a variety of venues in New York. He has written, produced
and directed a number of podcasts.

www.ingramcontent.com/pod-product-compliance
Lightning Source LLC
Chambersburg PA
CBHW040848210326
41597CB00029B/4767